Vol 1 reviewed in

Aust JP   1975   p. 276   Errol Martin (ANU, 1977)

Bib de la Phil   21   1974   p. 318   item 959
                                    (Phil Dept, Michigan, 1977)
J. Phil   1976   p. 149   Harold Hodes (Harvard)

Mind   1976   p. 129   Wilfrid Hodges (Bedford Cole)

Phil Rev   1976   p. 102   Stanley C Martens (Cornell)

Times Lit Supp.   May 24   1974   p. 567
    (with Hao Wang From M to Phil)   anon.

for 'proposition' ("I shall not attempt here " to
explain the notion of a proposition — p 46 of Vol 1)
see Vol 1 pp 46-7 ("I shall preserve a
distinction between a proposition and a
propositional expression " — p. 47) (But
"not necessarily insisting that such a distinction
must be drawn, for my intention is to remain
neutral between all accounts of proposition
which satisfy certain minimum requirements," v12
"every prop. is either true or false " and none is both, and
for all propositions p there is something, v12: the proposition
that p, that is true if and only if p.

LOGIC AND ARITHMETIC

Logic and Arithmetic

** Rational and Irrational Numbers

DAVID BOSTOCK

OXFORD · AT THE CLARENDON PRESS
1979

Oxford University Press, Walton Street, Oxford OX2 6DP

OXFORD LONDON GLASGOW
NEW YORK TORONTO MELBOURNE WELLINGTON
KUALA LUMPUR SINGAPORE JAKARTA HONG KONG TOKYO
DELHI BOMBAY CALCUTTA MADRAS KARACHI
NAIROBI DAR ES SALAAM CAPE TOWN

British Library Cataloguing in Publication Data

Bostock, David
  Logic and arithmetic.
  2: Rational and irrational numbers.
  1. Logic, Symbolic and mathematical
  2. Axioms   3. Arithmetic
  513'.01   QA9   79–40414

  ISBN 0–19–824591–2

Typeset by Cotswold Typesetting Ltd., Gloucester
Printed in Great Britain at the University Press, Oxford
by Eric Buckley, Printer to the University

Preface

MUCH of the first volume of this work was devoted more to logic than to arithmetic, and I made some attempt there to defend the classical, two-valued, impredicative, higher-order logic. As I tried to explain, it does not seem to me that this logic requires an *ontological* backing, and my attitude to logic is thus classical without being Platonistic. Roughly, my strategy was to argue that the analysis of propositions into components of various types, and the use of quantified variables in place of those components, need not be taken as involving any assumption of the independent existence of those components, since the whole procedure can be understood syncategorematically. So in logic, as I see it, there need be no commitment to any entities, save possibly to propositions themselves. I offered no further account of propositions there, beyond remarking that one might give an analysis in terms of *possible* sayings or thinkings, but again my preference would be to rely on the notion of possibility required for such an analysis rather than on an existential postulate which I see no hope of arguing for. I persist, therefore, in a non-Platonic attitude to logic, while at the same time feeling free to make full use of classical (impredicative) logic in discussing arithmetic. Indeed my underlying logic goes beyond what can reasonably be called 'classical', for against the background of a higher-order system bound by the restrictions of simple type theory I also employ 'type-neutral' elements such as the pure quantifiers. The logic of these quantifiers was introduced in the first volume (Chapter 4), but the present volume extends the use of type-neutrality to other kinds of components with very little by way of further justification.

There is obviously more to be said on all of these topics, but I say no more of them here. The bulk of the present volume was written during a sabbatical leave in 1975, and since then I have attempted to clarify my views on logic in two ways. I have thought further on type-neutrality, and I hope to be publishing a brief monograph on it shortly. In this I study type-neutral elements of various kinds, with no limitation to 'quantifiers', and I attempt to present the subject with greater formal clarity than in Chapter 4 of the first volume. In fact I regard Chapter 4 as being superseded by the new treatment, since it will contain an important correction to the system put forward there. In another direction, I have tried to take further the debate on the foundations of classical logic, setting my views in the context of a discussion of Frege and Russell and their successors. However, all this lies elsewhere. As far as the present work is concerned I say no more on logic, and the present volume aims only to continue the discussion of number and to bring it to a conclusion.

I begin in Chapter 1 by recalling the main outline of the logicist construction of natural numbers that was put forward in the first volume and offering a variation suited to cardinal numbers rather than counting numbers. Then I show how the basic idea may also be adapted to yield yet more logicist constructions of natural numbers based upon further uses or applications of those numbers. This sets the stage for what I see as the chief philosophical problem for logicist (or more generally reductionist) accounts of number—the great variety of ways in which such reductions can apparently be achieved. Of course it is well known that the common identification of numbers with sets is open to objection on this score, but that identification seems to me to be lacking in intuitive appeal anyway, and I began by dismissing it (Volume 1, Chapter 1, §1). What is interesting is that the more radically nominalist reductions which refuse to treat numbers as objects of any sort, are still subject to a very similar objection. (I do not imagine that this situation is peculiar to nominalist reductions of *number* in particular, but it is specially clear in this case because of the relatively sharp formal requirements which these reductions must satisfy.) My suggestions as to how the objection can best be met, and what significance the reductions in fact have, are postponed to Chapter 4. Meanwhile Chapters 2 and 3 widen the enquiry by looking into the availability of comparable reductions for rational and for real numbers.

Classically rational and real numbers are reduced to natural numbers, but the reduction I am mainly concerned with is not of this sort but resembles reductions for natural numbers in working directly from the *application* of rational and of real numbers. This I take to lie chiefly in the field of measurement, but it is necessary to begin with some discussion of this point since contemporary accounts of measurement are mostly written from a rather different viewpoint.[1] My reflections here lead me to take the notion of a *part* as fundamental, and the logical construction begins (Chapter 2, §4) with axioms first for the general notion of a part and then for the special case of 'additive' parts. My general axioms for parts make less sweeping existential demands than the well-known system of 'mereology' or 'the calculus of individuals', and my axiom for additive parts is again weaker than usual in not being essentially restricted to finite wholes. Partly for this reason and partly because it is anyway unusual for this topic to be approached formally in terms of parts, I had designed (and written) a formal develop-ment of the resulting system to give some indication of its potentialities and to provide proofs of theorems cited in the main text. However, this design has had to be abandoned for economic reasons: the printing of logical symbols is costly and this book is already far too expensive. Because of this change of plan the logical development remains somewhat fragmentary, and many theorems are cited without any indication of their proof. I must apologize here for this.

I next show (Chapter 2, §6) how the more traditional postu-lates for measurable quantities may be deduced from the basis I provide, with the exception that my existential assumptions are again weaker than usual. It turns out that this weakness is of no noticeable importance to the resulting theory and virtually all the traditional theorems are still available, but again I have had to excise my detailed proof of this assertion. The theory in question is essentially the Greek theory of proportion (which I think should be better known than it is) and it is from this basis

---

[1] The treatment of what is called 'measurement' in Russell and Whitehead's *Principia mathematica* is pursued from a viewpoint which has something in common with mine, but I do not discuss it since they are in fact treating of a quite different subject. They are wholly concerned with *vectors*, such as 'two miles due north', which are of course relevant in geometry but have no appli-cation to, say, the measurement of mass. (I treat of distances, among other things, but not of directions.)

that I suggest that the theory of rational numbers, and later of real numbers, might be constructed. Some difficulties of detail do emerge, and I think that they shed further light on what, in general, such reductions can achieve, but in the main it turns out that rational and real numbers can be treated in fairly close parallel to natural numbers.

Other varieties of number are mentioned only in a cursory way, and, apart from a small excursus on denumerability (Chapter 1, §4), I have nothing to say in this book about infinite numbers. However, I hope that my discussion has ranged widely enough to cast some light on the general concept of a number and the several strands to be discerned within it. I conclude the work with an attempt to draw the various strands together and to place the concept within an orthodox philosophical framework.

In writing this book I have been helped by discussion with many friends, but I should like in particular to record my debt to Professor Dana Scott. The logical construction in Chapters 2 and 3 has been much improved by what I have learned from him in discussion, at lectures, and from an unpublished treatment of the Greek theory of proportion which he has been good enough to show me. But the basic idea, of course, I have simply taken from Euclid.

DAVID BOSTOCK

Merton College, Oxford
January 1978

Contents

1. More on Natural Numbers

1. Finite Cardinals

IN the previous volume we considered some uses of natural numbers which are expressed in English by the numerals 'one', 'two', 'three', . . .—uses which relate to the procedure of counting, or more generally to 'how many?' questions. Our method was to concentrate entirely on the 'weak' numerical quantifiers expressed by 'there is at least one . . .', 'there are at least two . . .', and so on, and the first problem was to introduce these quantifiers. It was argued that these quantifiers occurred in conjunction with a *counting relation* which was not always non-identity, though non-identity had a good claim to be regarded as the most fundamental counting relation since all other acceptable counting relations are included in it. It was then shown how one could introduce the weak numerical quantifiers based on non-identity as a special variety of type-neutral quantifiers, and in terms of these one could then go on to introduce weak numerical quantifiers based on other counting relations. Finally it was suggested that statements directly concerning natural numbers could be seen as abbreviations of generalizations concerning numerical quantifiers, for the Peano postulates characterizing the structure of the number series can be proved to hold for the weak numerical quantifiers, and this way of viewing statements directly concerning natural numbers evidently allows us to transfer the postulates from the numerical quantifiers to the numbers themselves. Essentially the strategy of this last step was to construe natural numbers as obtained by a sort of 'abstraction' from weak numerical quantifiers, except that the abstraction technique did not involve any axiom of abstraction or any identifying of numbers with other entities previously available. For the formulae of the theory of natural numbers were introduced *as wholes*, and nothing was provided to be what numerals referred to, or what numerical variables ranged over, in the sense in which that expression is usually understood.

The main difficulty which we saw with this approach to the natural numbers was its apparent arbitrariness; we were concentrating wholly upon the way numbers appear within weak numerical quantifiers and excluding from consideration all other contexts involving numbers. One very obvious alternative was to consider not the weak quantifiers 'there are at least n . . .' but the strong quantifiers 'there are exactly n . . .', and perhaps it will be of interest if we now look more closely into this alternative. The general lines of the construction will be closely similar to that given for weak quantifiers, and so may serve as a more detailed recapitulation of the essential points of that earlier construction, but there is one variation of some interest[1].

We must begin once more with the concept of a *pure* or *type-neutral* quantifier, capable of being significantly applied to any monadic propositional form, or, as we may also say, to any monadic predicate of any type. I use Q and q (and their alphabetic variants) as schematic letter and variable for such quantifiers, and these occur primarily in contexts of the form $(Q\alpha)(\Phi\alpha)$. The symbol α (and its alphabetic variants) I call a *dummy* variable, because unlike normal variables it is not intended as a variable for any particular type of propositional component but is used *only* to indicate the bondage links; it could be dropped from the symbolism without loss if we adopted some other method of indicating these links, for example by drawing arrows. The symbol Φ (and its alphabetic variants) is a schematic letter for anything that a pure quantifier may be applied to, i.e. for a monadic propositional form or predicate of any type. Within its range are included the predicates of pure quantifiers themselves (or the monadic propositional forms written with a pure quantifier letter as their schematic letter), and when we wish to restrict our attention to these we write Φ immediately followed by a pure quantifier letter (or variable) and not by a mere dummy variable α. Thus as a special case of the schema $(Q\alpha)(\Phi\alpha)$ we have $(Qq)(\Phi q)$, so that (where ϕ is

---

[1] I regard the present construction as an improvement on the one very briefly indicated in footnote 15 on p. 209 of volume 1. There is no need to introduce one–one correlations when we are confining attention to finite numbers.

a variable answering to the schematic letter Φ) we may assert

$$(\forall\phi)(\ldots(Q\alpha)(\phi\alpha)\ldots)\rightarrow(\ldots(Qq)(\Phi q)\ldots).$$

This type of inference is essential for the ensuing deduction.

The first step in the construction is to single out from among the pure quantifiers the strong numerical quantifiers based on non-identity as counting relation. To this end we define

$$(0\alpha)(\Phi\alpha) \quad \text{for} \quad \sim(\exists\alpha)(\Phi\alpha)$$

$$(Q'\alpha)(\Phi\alpha) \quad \text{for} \quad (\exists\alpha)(\Phi\alpha \ \& \ (Q\beta)(\Phi\beta \ \& \ \beta\neq\alpha))$$

$$Q_1\equiv Q_2 \quad \text{for} \quad (\forall\phi)((Q_1\alpha)(\phi\alpha)\leftrightarrow(Q_2\alpha)(\phi\alpha))$$

$$Q_1 S Q_2 \quad \text{for} \quad (\forall\phi)((Q_1\alpha)(\phi\alpha)\leftrightarrow(Q_2'\alpha)(\phi\alpha))$$

The identity occurring in the definition of Q' is Leibnizian, and therefore type-neutral, so both $(0\alpha)(-\alpha-)$ and $(Q'\alpha)(-\alpha-)$ are type-neutral quantifiers, i.e.

$$(\exists q)(q\equiv 0)$$

$$(\exists q)(q\equiv Q').$$

We now introduce the ancestral $*S$ of the relation S and confine our attention to the pure quantifiers which bear the ancestral of this relation to the quantifier $(0\alpha)(-\alpha-)$. These are the strong numerical quantifiers based on non-identity, so we define

$$(\forall n)(\ldots(n\alpha)(-\alpha-)\ldots) \quad \text{for} \quad (\forall q)(q*S0\rightarrow(\ldots(q\alpha)(-\alpha-)\ldots))$$

$$(\exists n)(\ldots(n\alpha)(-\alpha-)\ldots) \quad \text{for} \quad (\exists q)(q*S0 \ \& \ (\ldots(q\alpha)(-\alpha-)\ldots))$$

From the properties of the ancestral it is then easy to show both that mathematical induction applies to our numerical quantifiers and that

$$(\exists n)(n\equiv 0)$$

$$(\exists n)(n\equiv m').$$

We also have, as immediate consequences of our definitions,

$$0\not\equiv n'$$

$$n\equiv m \rightarrow n'\equiv m'.$$

To obtain Peano's postulates, then, what is required is just the converse of this last theorem, i.e.

$$n' \equiv m' \rightarrow n \equiv m.$$

so we may now turn to the task of deducing this.

First we may note that the properties of the ancestral at once yield the theorem

(1) $\qquad\qquad n \equiv 0 \lor (\exists m)(n \equiv m').$

Next we introduce an auxiliary definition

$$n \geqslant m$$

for

$$(\forall\phi)((n\alpha)(\phi\alpha) \rightarrow (\exists\psi)((m\alpha)(\psi\alpha)\ \&\ (\forall\alpha)(\psi\alpha \rightarrow \phi\alpha)))$$

and we note as more or less immediate consequences of this definition

(2) $n \equiv m \rightarrow n \geqslant m$

(3) $n \geqslant m\ \&\ m \geqslant p \rightarrow n \geqslant p$

(4) $n' \geqslant n$

(5) $n \geqslant 0$

(6)[2] $0 \geqslant n \rightarrow 0 \equiv n$

(7) $\sim(\exists\psi)(m\alpha)(\psi\alpha) \rightarrow (n \geqslant m \rightarrow n \equiv m).$

To obtain more interesting results we shall also need two lemmas which are parallel to the lemmas of Chapter 1 of Volume 1 and proved by induction in the same way, namely

Lem (i) $(\forall\alpha)(\Phi\alpha \leftrightarrow \Psi\alpha) \rightarrow ((n\alpha)(\Phi\alpha) \leftrightarrow (n\alpha)(\Psi\alpha))$

Lem (ii) $\Phi\beta\ \&\ (n'\alpha)(\Phi\alpha) \rightarrow (n\alpha)(\Phi\alpha\ \&\ \alpha \neq \beta)$

Now using the definition of $\geqslant$ and lemma (i) it is quite straightforward to show that

(8) $n \geqslant m \rightarrow n' \geqslant m'.$

We can also establish a qualified converse to this, namely

(9) $(\exists\psi)(m\alpha)(\psi\alpha) \rightarrow (n' \geqslant m' \rightarrow n \geqslant m).$

[2] $0 \not\geqslant n'$ is immediate from the definition. Then use theorem (1).

The argument is as follows. We assume the antecedents

   (i) $(\exists\psi)(m\alpha)(\psi\alpha)$

   (ii) $n' \geqslant m'$

and in addition

   (iii) $(n\alpha)(\Phi\alpha)$.

Now in view of (i) we clearly have

   (iv) $(\forall\alpha)(\Phi\alpha) \rightarrow (\exists\psi)((m\alpha)(\psi\alpha) \ \& \ (\forall\alpha)(\psi\alpha \rightarrow \Phi\alpha))$.

So we next assume that for some β

   (v) $\sim\Phi\beta$.

From (iii), (v), and lemma (i) it then follows that

   $(n'\alpha)(\Phi\alpha \lor \alpha = \beta)$

and hence, using (ii), that for some Ψ'

   $(m'\alpha)(\Psi'\alpha) \ \& \ (\forall\alpha)(\Psi'\alpha \rightarrow \Phi\alpha \lor \alpha = \beta)$.

We now consider two cases according as $\Psi'\beta$ or $\sim\Psi'\beta$. In the first case we have, using lemma (ii),

   $(m\alpha)(\Psi'\alpha \ \& \ \alpha \neq \beta) \ \& \ (\forall\alpha)(\Psi'\alpha \ \& \ \alpha \neq \beta \rightarrow \Phi\alpha)$

and therefore

   $(\exists\psi)((m\alpha)(\psi\alpha) \ \& \ (\forall\alpha)(\psi\alpha \rightarrow \Phi\alpha))$

In the second case we have

   $(m'\alpha)(\Psi'\alpha) \ \& \ (\forall\alpha)(\Psi'\alpha \rightarrow \Phi\alpha)$

and therefore

   $(\exists\gamma)(\Psi'\gamma \ \& \ (m\alpha)(\Psi'\alpha \ \& \ \alpha \neq \gamma) \ \& \ (\forall\alpha)(\Psi'\alpha \ \& \ \alpha \neq \gamma \rightarrow \Phi\alpha))$.

So once again

   $(\exists\psi)((m\alpha)(\psi\alpha) \ \& \ (\forall\alpha)(\psi\alpha \rightarrow \Phi\alpha))$.

Now we have only to assemble the pieces. By conditionalizing and generalizing the argument from assumption (v), we infer that

   $(\exists\alpha)(\sim\Phi\alpha) \rightarrow (\exists\psi)((m\alpha)(\psi\alpha) \ \& \ (\forall\alpha)(\psi\alpha \rightarrow \Phi\alpha))$.

Putting this conclusion together with (iv) and conditionalizing the argument from assumption (iii) we therefore have

$$(n\alpha)(\Phi\alpha) \rightarrow (\exists\psi)((m\alpha)(\psi\alpha) \ \& \ (\forall\alpha)(\psi\alpha \rightarrow \Phi\alpha))$$

Generalizing once more we obtain the required conclusion

$$n \geqslant m$$

and this completes the proof. To theorems (8) and (9) we may now add

(10) $n \geqslant m \ \& \ m \geqslant n \rightarrow n \equiv m$

for this is now easily established by induction. The induction basis is given by theorem (6), and the induction step is easily proved by considering cases and applying either theorem (7) or theorem (9) as appropriate.

From the position we have reached so far it is clear that the proof of Peano's postulates will be completed if we can only remove the qualification from theorem (9) by establishing unconditionally that

$$(\exists\psi)(m\alpha)(\psi\alpha)$$

and this we can do by proving Frege's theorem. As a preliminary we establish two more results about the relation $\geqslant$. First

(11) $n' \geqslant m \ \& \ n' \not\equiv m \rightarrow n \geqslant m$.

This is proved by induction on m. The induction basis is given by theorem (5), and the induction step may be argued thus. Assume

(i) $n' \geqslant m' \ \& \ n' \not\equiv m'$.

Then by theorem (7) we have

$$(\exists\psi)(m'\alpha)(\psi\alpha)$$

and by theorem (7) again, with theorem (4), this implies

$$(\exists\psi)(m\alpha)(\psi\alpha)$$

so theorem (9) applies, and we infer

(ii) $n \geqslant m$.

Also, in view of the assumption $n' \not\equiv m'$ we have

(iii) $n \not\equiv m$.

However, from (ii) and (iii) we infer, using theorems (6) and (1), that for some p

(iv) $n \equiv p'$

Now putting (ii), (iii), and (iv) together we have

$$p' \geqslant m \ \& \ p' \not\equiv m$$

so by the hypothesis of induction

$$p \geqslant m$$

and therefore by theorem (8)

$$p' \geqslant m'$$

i.e., using (iv) again,

$$n \geqslant m'.$$

This completes the induction step. Finally we elaborate this last theorem to

(12) $n \geqslant m \ \& \ n \not\equiv m \leftrightarrow n' \geqslant m \ \& \ n' \not\equiv m \ \& \ n \not\equiv m.$

The implication from right to left is given by theorem (11), and the implication from left to right depends only on theorems (3), (4), and (10).

We may now turn to the derivation of Frege's theorem. This theorem requires us to count the numerical quantifiers themselves, so we note first that in view of the type-neutrality of these quantifiers formulae of the form $(nq)(\Phi q)$ are already available to us as special cases of formulae $(n\alpha)(\Phi\alpha)$. However, the formulae $(nq)(\Phi q)$ would be used to count quantifiers under the relation of non-identity whereas it is clear that we shall need to count them under the relation of non-equivalence, and without an axiom of extensionality it cannot be assumed that these relations coincide. We must therefore make use of strong numerical quantifiers with counting relations other than non-identity, but it turns out that this notion is altogether less clear than it was in the case of the weak numerical quantifiers. As a temporary measure, therefore, I here avoid a general definition and give one that applies only to the counting relation of non-equivalence between quantifiers, *viz.*

$$(n_{\neq}q)(\Phi q)$$

for

$$(\exists\psi)((nq)(\psi q) \ \& \ (\forall q_1 q_2)(\psi q_1 \ \& \ \psi q_2 \ \& \ q_1 \neq q_2 \rightarrow q_1 \not\equiv q_2)$$

$$\& \ (\forall q_1)(\psi q_1 \rightarrow \Phi q_1) \ \& \ (\forall q_1)(\Phi q_1 \rightarrow (\exists q_2)(\psi q_2 \ \& \ q_1 \equiv q_2))).$$

For the particular application that we require we may also define further

$$(n_{\neq}m)(\Phi m) \quad \text{for} \quad (n_{\neq}q)(q*\mathbf{S}0 \ \& \ \Phi q).$$

Now from these definitions it is easy to see that

(13) $\quad n \equiv m \rightarrow ((n_{\neq}p)(\Phi p) \leftrightarrow (m_{\neq}p)(\Phi p)).$

Also we may in a straightforward way prove the recursive equivalences

(14) $\quad (0_{\neq}m)(\Phi m) \leftrightarrow \sim(\exists m)(\Phi m)$

$$(n'_{\neq}m)(\Phi m) \leftrightarrow (\exists m)(\Phi m \ \& \ (n_{\neq}p)(\Phi p \ \& \ p \not\equiv m)).$$

These in turn enable us to re-establish lemma (i) in a form appropriate for the new numerical quantifiers, *viz.*

Lem (i′) $\quad (\forall m)(\Phi m \leftrightarrow \Psi m) \rightarrow ((n_{\neq}m)(\Phi m) \leftrightarrow (n_{\neq}m)(\Psi m))$

The proof of Frege's theorem now presents no difficulty. In view of theorems (6) and (14) we evidently have

$$(0_{\neq}m)(0 \geqslant m \ \& \ 0 \not\equiv m)$$

which will serve as the basis for an inductive argument. Next we assume as the hypothesis of induction

$$(n_{\neq}m)(n \geqslant m \ \& \ n \not\equiv m)$$

and we consider two cases according as $n \equiv n'$ or $n \not\equiv n'$. In the first case we have by lemma (i′)

$$(n_{\neq}m)(n' \geqslant m \ \& \ n' \not\equiv m)$$

and hence by theorem (13)

$$(n'_{\neq}m)(n' \geqslant m \ \& \ n' \not\equiv m)$$

as desired. In the second case we have by theorem (12) and lemma (i′)

$$(n_{\neq}m)(n' \geqslant m \ \& \ n' \not\equiv m \ \& \ n \not\equiv m)$$

but also by theorem (4) and the case hypothesis

$$n' \geqslant n \ \& \ n' \not\equiv n.$$

Putting these together and applying existential generalization we have

$$(\exists p)(n' \geqslant p \ \& \ n' \not\equiv p \ \& \ (n \not\equiv m)(n' \geqslant m \ \& \ n' \not\equiv m \ \& \ p \not\equiv m))$$

and so by theorem (14)

$$(n' \not\equiv m)(n' \geqslant m \ \& \ n' \not\equiv m).$$

This completes the induction. We have established Frege's theorem in the form

$$(n \not\equiv m)(n \geqslant m \ \& \ n \not\equiv m).$$

In view of the definition of $(n \not\equiv m)(-\!-m-\!-)$ it evidently follows that

$$(\exists \psi)(n\mathrm{q})(\psi\mathrm{q})$$

and so finally

$$(\exists \psi)(n\alpha)(\psi\alpha).$$

The proof of Peano's postulates is therefore complete.

Let us now turn to discuss the point left undecided in this construction, namely the question of giving a general definition of strong numerical quantifiers with counting relations other than non-identity. In the case of the weak quantifiers we offered the definition (Vol. 1, p. 151).

$$(\exists n_{\mathfrak{R}}\alpha)(\Phi\alpha)$$

for

$$(\exists \psi)((\exists n\alpha)(\psi\alpha) \ \& \ (\forall\alpha\beta)(\psi\alpha \ \& \ \psi\beta \ \& \ \alpha \neq \beta \rightarrow \alpha\mathfrak{R}\beta)$$
$$\& \ (\forall\alpha)(\psi\alpha \rightarrow \Phi\alpha)).$$

This definition was reached by generalizing from an analysis of a particular statement of the form $(\exists n_{\mathfrak{R}}\alpha)(\Phi\alpha)$, and then further supported by showing that where the counting relation $\mathfrak{R}$ obeys the minimum condition for an acceptable counting relation, *viz.* where it is irreflexive and symmetrical, we can deduce from the definition the expected recursive equivalences

$$(\exists 1_{\mathfrak{R}}\alpha)(\Phi\alpha) \leftrightarrow (\exists\alpha)(\Phi\alpha)$$
$$(\exists n'_{\mathfrak{R}}\alpha)(\Phi\alpha) \leftrightarrow (\exists\alpha)(\Phi\alpha \ \& \ (\exists n_{\mathfrak{R}}\beta)(\Phi\beta \ \& \ \beta\mathfrak{R}\alpha)).$$

It is easily checked that the direction of the argument could have been reversed; we might instead have started from the assumption that the recursive equivalences must hold at least for irreflexive and symmetrical relations, and then deduced the (equivalence corresponding to) the definition directly from this assumption.

Suppose then that we attempt the reverse direction of argument with the strong quantifiers. We shall evidently expect the appropriate recursive equivalences

$$(0_\Re\alpha)(\Phi\alpha) \leftrightarrow \sim(\exists\alpha)(\Phi\alpha)$$

$$(n'_\Re\alpha)(\Phi\alpha) \leftrightarrow (\exists\alpha)(\Phi\alpha \ \& \ (n_\Re\beta)(\Phi\beta \ \& \ \beta\Re\alpha))$$

to hold for all acceptable countable relations, and we may begin by assuming that in this case too an acceptable counting relation is one that is irreflexive and symmetrical. Now it is quite simple to prove (by induction) that if these recursive equivalences always hold[3] where $\Re$ is irreflexive and symmetrical then also where $\Re$ is irreflexive and symmetrical the formula

$$(n_\Re\alpha)(\Phi\alpha)$$

will always be equivalent to

$$(\exists\psi)((n\alpha)(\psi\alpha) \ \& \ (\forall\alpha\beta)(\psi\alpha \ \& \ \psi\beta \ \& \ \alpha\neq\beta \rightarrow \alpha\Re\beta)$$

$$\& \ (\forall\alpha)(\psi\alpha\rightarrow\Phi\alpha) \ \& \ (\forall\alpha)(\Phi\alpha\rightarrow(\exists\beta)(\psi\beta \ \& \ \sim\alpha\Re\beta))).$$

Clearly the definition thus suggested is simply a generalization of the definition given earlier for the particular case of the counting relation of non-equivalence between quantifiers. However, at first sight it would appear that in this case the method of argument must have led us astray.

The difficulty emerges when we turn to consider the standard connection between strong and weak numerical quantifiers, *viz.*

$$(n_\Re\alpha)(\Phi\alpha) \leftrightarrow (\exists n_\Re\alpha)(\Phi\alpha) \ \& \ \sim(\exists n'_\Re\alpha)(\Phi\alpha)$$

(except where $n \equiv 0$, for then the right-hand side is undefined).

---

[3] I mean that they hold *for all* Φ. This is a stronger assumption than we need. Using the terminology introduced on p. 15, a sufficient assumption would be that Φ satisfies the condition

$$(\forall\psi)(\psi\text{-}restricts_\Re\Phi\rightarrow(((0_\Re\alpha)(\psi\alpha)\leftrightarrow\sim(\exists\alpha)(\psi\alpha))$$

$$\& \ (\forall n)((n'_\Re\alpha)(\psi\alpha)\leftrightarrow(\exists\alpha)(\psi\alpha \ \& \ (n_\Re\beta)(\psi\beta \ \& \ \beta\Re\alpha)))))$$

Now we find that if we simply make the assumption that $\mathfrak{R}$ satisfies the recursive equivalences for both strong and weak quantifiers[4] then this by itself allows us to prove (by induction)

$$(\exists n_{\mathfrak{R}}\alpha)(\Phi\alpha) \;\&\; \sim(\exists n'_{\mathfrak{R}}\alpha)(\Phi\alpha) \to (n_{\mathfrak{R}}\alpha)(\Phi\alpha)$$

which is the more doubtful half of the connecting thesis, for we should surely require that any adequate definition of $n_{\mathfrak{R}}$ must yield, for acceptable counting relations $\mathfrak{R}$,

$$(n_{\mathfrak{R}}\alpha)(\Phi\alpha) \to (\exists n_{\mathfrak{R}}\alpha)(\Phi\alpha)$$

$$(n_{\mathfrak{R}}\alpha)(\Phi\alpha) \to \sim(\exists n'_{\mathfrak{R}}\alpha)(\Phi\alpha).$$

Actually the first of these is again forthcoming from the simple assumption that $\mathfrak{R}$ satisfies all the recursive equivalences, but the second requires in addition the assumption[5]

$$\Phi\beta \;\&\; (\exists n'_{\mathfrak{R}}\alpha)(\Phi\alpha) \to (\exists n_{\mathfrak{R}}\alpha)(\Phi\alpha \;\&\; \alpha\mathfrak{R}\beta)$$

which is much stronger than the mere assumption that $\mathfrak{R}$ is irreflexive and symmetrical. However, since we should anyway want to insist that 'there are exactly n' and 'there are more than n' cannot be true together, this line of argument leads to the conclusion that the strong quantifiers may adequately be defined in terms of the weak *via* this connecting thesis. We can of course frame the definition wholly in terms of strong quantifiers if we wish, since the weak quantifiers may themselves be defined in terms of the strong by using the equivalence[6]

$$(\exists n\alpha)(\Phi\alpha) \leftrightarrow (\exists\psi)((n\alpha)(\psi\alpha) \;\&\; (\forall\alpha)(\psi\alpha\to\Phi\alpha))$$

or more generally

$$(\exists n_{\mathfrak{R}}\alpha)(\Phi\alpha) \leftrightarrow (\exists\psi)((n\alpha)(\psi\alpha) \;\&\; (\forall\alpha\beta)(\psi\alpha \;\&\; \psi\beta \;\&\; \alpha\neq\beta \to \alpha\mathfrak{R}\beta)$$
$$\&\; (\forall\alpha)(\psi\alpha\to\Phi\alpha)).$$

So our second line of argument leads us to the definition

---

[4] Note 3 still applies.

[5] Note 3 applies once more. The appropriate formulation is given on p. 15.

[6] If we make no exception for $n \equiv 0$ then $(\exists 0_{\mathfrak{R}}\alpha)(\Phi\alpha)$ is defined so as to be trivially true for every Φ.

$$(n_{\Re}\alpha)(\Phi\alpha)$$

for

$$(\exists\psi)((n\alpha)(\psi\alpha) \;\&\; (\forall\alpha\beta)(\psi\alpha \;\&\; \psi\beta \;\&\; \alpha \neq \beta \;\rightarrow\; \alpha\Re\beta) \;\&\; (\forall\alpha)(\psi\alpha{\rightarrow}\Phi\alpha))$$

$$\&\; \sim(\exists\psi)((n'\alpha)(\psi\alpha) \;\&\; (\forall\alpha\beta)(\psi\alpha \;\&\; \psi\beta \;\&\; \alpha \neq \beta \;\rightarrow\; \alpha\Re\beta)$$

$$\&\; (\forall\alpha)(\psi\alpha{\rightarrow}\Phi\alpha)).$$

These two definitions do not in general coincide. In order to deduce the first from the second we must assume that $\Re$ is irreflexive and symmetrical, and in order to deduce the second from the first we must again assume that $\Re$ satisfies the thesis

$$\Phi\beta \;\&\; (\exists n'_{\Re}\alpha)(\Phi\alpha){\rightarrow}(\exists n_{\Re}\alpha)(\Phi\alpha \;\&\; \alpha\Re\beta).$$

Where a relation $\Re$ does not satisfy this thesis it is easy to see that the first definition suggested would yield absurd results. For example the relation of being tangent to (between circles) is irreflexive and symmetrical but does not satisfy the thesis, as we see by noticing that in this diagram there are at least three

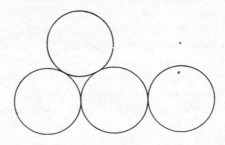

tangent circles (namely the leftmost three) but it is not true that for each circle in the diagram there are at least two tangent circles which are tangent to it (the rightmost circle being a counter-example). Accordingly, if we apply our first definition to this situation we find, as might have been expected, that there are in the diagram exactly three tangent circles, but *also* that there are in the diagram exactly *two* tangent circles (the verifying pair being the two rightmost circles). This is obviously a ridiculous consequence, and it seems to show that the first definition should be rejected. The second definition evidently cannot be made to yield such apparently contradictory results, but it suffers from the fact that it will not yield the recursive equivalences without further assumption. For example in the diagram there is a circle (namely the rightmost) such that there is exactly one circle in the diagram tangent to it, but according to the second definition it is not true that there are exactly two tangent circles in the diagram. So here we have a counter-example to the principle

$$(\exists\alpha)(\Phi\alpha \ \& \ (1_{\mathfrak{R}}\beta)(\Phi\beta \ \& \ \beta\mathfrak{R}\alpha)) \to (2_{\mathfrak{R}}\alpha)(\Phi\alpha).$$

Since it seems to me that the recursive equivalences are quite fundamental to our idea of a numerical quantifier, this is something of a difficulty.

The most obvious remedy is to strengthen the criteria for being an acceptable counting relation. Where the strong numerical quantifiers are concerned we might say that a relation is an acceptable counting relation only if it is irreflexive and symmetrical *and* satisfies the crucial thesis

$$\Phi\beta \ \& \ (\exists n'_{\mathfrak{R}}\alpha)(\Phi\alpha) \to (\exists n_{\mathfrak{R}}\alpha)(\Phi\alpha \ \& \ \alpha\mathfrak{R}\beta).$$

It turns out that this revised criterion reduces to something much simpler, for if a relation $\mathfrak{R}$ satisfies this thesis *for all* Φ then in particular

$$\sim\beta\mathfrak{R}\beta \ \& \ (\exists 2_{\mathfrak{R}}\alpha)(\sim\alpha\mathfrak{R}\beta) \to (\exists 1_{\mathfrak{R}}\alpha)(\sim\alpha\mathfrak{R}\beta \ \& \ \alpha\mathfrak{R}\beta)$$

whence

$$\sim\beta\mathfrak{R}\beta \to \sim(\exists 2_{\mathfrak{R}}\alpha)(\sim\alpha\mathfrak{R}\beta).$$

Assuming that $\mathfrak{R}$ is irreflexive and symmetrical this implies

$$\sim(\exists\alpha\gamma)(\sim\alpha\mathfrak{R}\beta \ \& \ \sim\beta\mathfrak{R}\gamma \ \& \ \alpha\mathfrak{R}\gamma)$$

which is to say that the negation of $\mathfrak{R}$ is transitive. Hence a relation can satisfy all three conditions only if it is the negation of an equivalence relation, and conversely it is easily checked that any relation which is the negation of an equivalence relation must satisfy all three conditions. Therefore the revised criterion is simply that, where the strong numerical quantifiers are concerned, the *acceptable* counting relations are just those that are negations of equivalence relations. With these counting relations our two suggested definitions will coincide, and the recursive equivalences will of course be forthcoming.

This is the simplest revision, but it might well be thought over-restrictive. After all where we wish to count the things α such that $\Phi\alpha$ under a particular counting relation $\Re$ it seems rather unnecessary to require that $\Re$ satisfy the thesis

$$\Phi\beta \ \& \ (\exists n'_{\Re}\alpha)(\Phi\alpha) \rightarrow (\exists n_{\Re}\alpha)(\Phi\alpha \ \& \ \alpha\Re\beta)$$

whatever we put for Φ, since it is only one particular value of Φ that immediately concerns us. For example if $\Re$ is taken as the relation between classes of being mutually exclusive, then although $\Re$ does not always satisfy the thesis it certainly does in the case where Φ is chosen so that $\Phi\alpha$ is of the form

$$(nx)(x\in\alpha) \ \& \ (\forall x)(x\in\alpha\rightarrow Fx).$$

Further, for such a choice of Φ the two definitions of $(n_{\Re}\alpha)(\Phi\alpha)$ do coincide, and no difficulty arises over the recursive equivalences. What this suggests is that we should *relativize* the notion of an acceptable counting relation by replacing it with a criterion for when a relation $\Re$ should be *acceptable for counting the* Φ's. Clearly we shall still require that $\Re$ be irreflexive and symmetrical, at least among Φ's. Also we shall need to ensure that $\Re$ and Φ together satisfy the crucial thesis

$$\Phi\beta \ \& \ (\exists n'_{\Re}\alpha)(\Phi\alpha) \rightarrow (\exists n_{\Re}\alpha)(\Phi\alpha \ \& \ \alpha\Re\beta).$$

However, this by itself will not be enough. If this were all that we required then by adopting the second definition we could certainly prove each instance of the recursive equivalences for $\Re$ and Φ, for example

$$(2_{\Re}\alpha)(\Phi\alpha)\leftrightarrow(\exists\beta)(\Phi\beta \ \& \ (1_{\Re}\alpha)(\Phi\alpha \ \& \ \alpha\Re\beta)).$$

However, we would not yet have any warrant to add

$$(1_{\Re}\alpha)(\Phi\alpha \ \& \ \alpha\Re\beta) \leftrightarrow (\exists\gamma)(\Phi\gamma \ \& \ \gamma\Re\beta \ \&$$
$$(0_{\Re}\alpha)(\Phi\alpha \ \& \ \alpha\Re\beta \ \& \ \alpha\Re\gamma))$$

so we could not complete the expansion of $(2_{\Re}\alpha)(\Phi\alpha)$ into

$$(\exists\beta)(\Phi\beta \ \& \ (\exists\gamma)(\Phi\gamma \ \& \ \gamma\Re\beta \ \& \ \sim(\exists\delta)(\Phi\delta \ \& \ \delta\Re\beta \ \& \ \delta\Re\gamma)))$$

The information contained in the recursive equivalences would thus be severely curtailed. As one might say, the difficulty is that if $\Re$ is to be acceptable for counting the Φ's then we shall also want $\Re$ to be acceptable for counting the Φ's that bear $\Re$ to a given Φ, and the Φ's that bear $\Re$ to two given Φ's, and so on, but we have not yet included these latter conditions. Let us then say that Ψ *restricts* Φ (under $\Re$) when the Ψ's are precisely the Φ's that bear $\Re$ to some given Φ, i.e.

$$\Psi \; restricts_{\Re} \; \Phi \quad \text{for} \quad (\exists\beta)(\Phi\beta \;\&\; (\forall\alpha)(\Psi'\alpha \leftrightarrow \Phi\alpha \;\&\; \alpha\Re\beta)).$$

Then we form the ancestral $*$-*restricts*$_{\Re}$ of this relation by the standard method, and the extra condition that we require for $\Re$ to be acceptable for counting the Φ's can be put as

$$(\forall\psi)(\psi \; *\text{-}restricts_{\Re} \; \Phi \rightarrow (\forall n)(\forall\beta)(\psi\beta \;\&\; (\exists n'_{\Re}\alpha)(\psi\alpha)$$
$$\rightarrow (\exists n_{\Re}\alpha)(\psi\alpha \;\&\; \alpha\Re\beta))).$$

With this condition added we find once more that if $\Re$ is a relation acceptable for counting the Φ's then (a) the two suggested definitions of $(n_{\Re}\alpha)(\Phi\alpha)$ must coincide and (b) to adopt either of these definitions is equivalent to stipulating that for each specific numeral in place of n the formula $(n_{\Re}\alpha)(\Phi\alpha)$ is equivalent to its full recursive expansion. The relativized criterion of acceptability is rather more complex than our original 'absolute' criterion, but its effect is just as satisfactory.

However, a further complication now suggests itself. There would be little point in liberalizing our criteria so as to permit exclusiveness as a counting relation (for appropriate choices of Φ) if in fact the definition suggested still gave unintuitive results with this counting relation, and it could be argued that it does. For example consider the statement:

> The class of men in this room has exactly three mutually exclusive subclasses of four members each.

Using 'α' temporarily as a class variable, '$\alpha \; excl \; \beta$' for

$$\text{`}\sim(\exists x)(x\in\alpha \;\&\; x\in\beta)\text{'}$$

and 'Fx' for 'x is a man in this room', this statement appears to be analysable as

$$(3_{excl}\alpha)((4x)(x\in\alpha) \;\&\; \alpha \subseteq \{x : Fx\})$$

i.e.

$$(3_{excl}\alpha)((4x)(x\in\alpha) \;\&\; (\forall x)(x\in\alpha\rightarrow Fx)).$$

However, according to the definitions suggested (or to the recursive expansion) the statement is on this analysis true not only

if there are 12 men in the room but also if there are 13, 14, or
15 men in the room, and it could reasonably be said that this is
not how we should ordinarily understand it. I think that there
is some force in the contention that we would normally take the
'exactly' to imply that any three mutually exclusive classes of
four members each were jointly *exhaustive* of the class of men
in the room.

We could frame a definition that would have this consequence, for example

$$(n_{\Re}\alpha)(\Phi\alpha)$$

for

$$(\exists\psi)((n\alpha)(\psi\alpha) \ \& \ (\forall\alpha\beta)(\psi\alpha \ \& \ \psi\beta \ \& \ \alpha \neq \beta \rightarrow \alpha\Re\beta) \ \& \ (\forall\alpha)(\psi\alpha \rightarrow \Phi\alpha)$$
$$\& \ (\forall\alpha)((\exists\beta)(\Phi\beta \ \& \ \sim\alpha\Re\beta) \rightarrow (\exists\beta)(\psi\beta \ \& \ \sim\alpha\Re\beta))).$$

When $\Re$ is taken as the negation of an equivalence relation this
definition coincides with our earlier suggestions, and where $\Re$ is
taken as exclusiveness between classes the last clause implies

$$\cup\{\alpha:\Phi\alpha\} \subseteq \cup\{\alpha:\psi\alpha\}$$

as desired. Nevertheless I do not in fact think that this revised
definition would be an improvement, and the main objection is
of course that the recursive equivalences would then not be
forthcoming for the quantifiers based on exclusiveness which we
were wanting to admit. If it is granted that the recursive equivalences must hold for all genuine numerical quantifiers and if
'there are *exactly* three mutually exclusive . . .' is taken to mean
'there are three mutually exclusive *and jointly exhaustive* . . .',
then it follows at once that this locution does not express a
genuine numerical quantifier. If this is admitted, then there is
evidently no call to tinker with our general definition of numerical quantifiers to make it fit this case, and we may put the
example on one side as irrelevant to our present concern[7]. If,
on the other hand, 'there are *exactly three* mutually exclusive . . .'

---

[7] The statement may be treated as interchangeable with 'the class of men
in this room is exactly three times as numerous as any class of four members'
The appropriate relation 'exactly three times as numerous as' is defined as a
special case of relations of the form $\approx_3$ on pp. 131–2. Of course recursive equivalences do hold for such relations, but they are not of the same sort as the
recursive equivalences appropriate to numerical quantifiers.

is taken to be equivalent to 'there are *at least three but not at least four* mutually exclusive . . .' then there is no strong objection to allowing it as a numerical quantifier (given a suitable choice of Φ). The only doubt is whether the locution would ever be understood this way.

Even when we are dealing with a counting relation which is the negation of an equivalence relation there is often room for doubt over how the corresponding English locution would be understood. Presumably the quantifier 'there are exactly *n* non-equivalent . . .' would always be understood in the sense I suggest, but even such an ordinary counting relation as 'is not the same height as' would seem to give rise to ambiguities. The statement

There are exactly three men in the room of different height

would be interpreted by our definition as being compatible with there being any number of men in the room so long as between them they exemplified three and only three heights. However, I would not be much surprised to find that the statement was more commonly understood as implying that there were just three men in the room altogether (and that they were all of different heights). Either interpretation of the English seems to me possible, but it *might* be thought that there is already some departure from our ordinary idiom in recognizing the existence of the quantifier 'there are exactly *n* . . . of different height' understood as I suggest. This point can be made much more strongly with locutions like 'there are exactly *n* mutually exclusive . . .', where the ordinary interpretation (if any) seems to me entirely doubtful, and I think that it is fairly clear that 'there are exactly *n* mutually tangent . . .' would in fact *always* be understood to mean 'there are exactly *n* . . . *and* they are all tangent to one another'. Of course it is perfectly in order to stipulate a different sense for these expressions, if that would be useful, but the stipulation has very little connection with ordinary usage, and with the last example at any rate the failure of the recursive equivalences seems to make the stipulation quite useless.

To sum up, I think that ordinary usage gives us no very firm guidance on strong numerical quantifiers with counting relations other than non-identity, and that is why I began this discussion by instead considering the implications that could be

Hodes (JPhil p.154): If there are only n objects, all numerical identity-statements between numbers greater then n are true on this definition

extracted from the recursive equivalences. It seems clear that if these equivalences are taken as our fundamental guide then the notion of an *acceptable* counting relation must be more restricted for the strong quantifiers than it was for the weak. If we relativize the relevant notion of acceptability, then we can if we wish admit more counting relations than those that are negations of equivalence relations, but I am inclined to think that there is in fact no point in doing so. On the other hand, the usefulness of strong quantifiers based on non-equivalence has already been demonstrated in our proof of Frege's theorem.

As a coda to this discussion of counting relations other than non-identity I should remark that since strong quantifiers in general are defined in terms of strong quantifiers based on non-identity, and since the Peano postulates have been shown to apply to the latter, they can now easily be transferred to the former, however exactly we choose our definition and our criterion for an acceptable counting relation. Therefore we may if we wish offer the theory of strong quantifiers based on non-identity as itself a construction of arithmetic, or we may proceed more circuitously by construing arithmetical statements as generalizations over all strong numerical quantifiers, whatever their counting relations. On the first plan the fundamental arithmetical statements 'n is the same number as m' and 'n is the next number after m' are simply identified with $n \equiv m$ and $n \equiv m'$, while on the second plan they are identified rather with

$$(\forall \Re)(\forall \phi)((n_{\Re}\alpha)(\phi\alpha) \leftrightarrow (m_{\Re}\alpha)(\phi\alpha))$$
$$(\forall \Re)(\forall \phi)((n_{\Re}\alpha)(\phi\alpha) \leftrightarrow (m'_{\Re}\alpha)(\phi\alpha)).$$

As I tried to suggest in Volume 1 (Chap. 6, pp. 206–208), the more circuitous route may possibly have some philosophical advantages, but the deduction of Peano's postulates for numbers is essentially the same in either case.

Here then is my first example of two distinct approaches to the natural numbers which may both be characterized as logicist in spirit, and which both construct the theory of natural numbers as a by-product of the theory of certain applications of those numbers. They differ just in selecting different applications to work with. I shall not comment further on this situation until more evidence is at hand, for much of what follows will be designed to provide yet further examples of alternative logicist constructions. However, before I turn to these further examples, I should like to add a little more on the constructions which employ numerical quantifiers.

In these constructions I have attempted only to prove Peano's postulates, which are framed in terms of identity and succession of numbers, since the well-known categoricity of these postulates shows that they provide an adequate basis for all properties of numbers which are definable in terms of identity and succession, and this includes all the properties occurring in elementary arithmetic. However, with the present approach to natural numbers there is of course no necessity to define all further properties in this way. Many of them can perfectly well be introduced at an earlier stage as properties of the numerical quantifiers. For example in each of our constructions we have already introduced a relation $\geqslant$ between numerical quantifiers, in the case of the weak quantifiers defining

$$\exists n \geqslant \exists m \quad \text{for} \quad (\forall \phi)((\exists n\alpha)(\phi\alpha) \rightarrow (\exists m\alpha)(\phi\alpha))$$

and in the case of the strong quantifiers defining

$$n \geqslant m \quad \text{for} \quad (\forall \phi)((n\alpha)(\phi\alpha) \rightarrow (\exists \psi)((m\alpha)(\psi\alpha) \& (\forall\alpha)(\psi\alpha \rightarrow \Phi\alpha))).$$

Once Peano's postulates are available for the numerical quantifiers it is perfectly easy to show that these relations, so defined, have the expected properties. Again we could introduce addition as an operation to be performed directly on numerical quantifiers, defining say

$$(\exists(n+m)\alpha)(\Phi\alpha) \quad \text{for} \quad (\exists\psi)((\exists n\alpha)(\Phi\alpha \& \psi\alpha) \& (\exists m\alpha)(\Phi\alpha \& \sim\psi\alpha))$$

and

$$((n+m)\alpha)(\Phi\alpha) \quad \text{for} \quad (\exists\psi)((n\alpha)(\Phi\alpha \& \psi\alpha) \& (m\alpha)(\Phi\alpha \& \sim\psi\alpha)).$$

From these definitions it is again perfectly straightforward to establish that the operation defined has the usually recursive properties of addition, i.e. (in the case of the weak quantifiers)

$$\exists(1+n) \equiv \exists n'$$

$$\exists(m'+n) \equiv \exists(m+n)'$$

and hence by the usual arguments we may obtain such fundamental properties of addition as

$$(\exists p)(\exists p \equiv \exists(n + m))$$

$$\exists(n + m) \equiv \exists(m + n)$$

$$\exists(n + (m + p)) \equiv \exists((n + m) + p)$$

$$(\exists n \equiv \exists m) \leftrightarrow (\exists(n + p) \equiv \exists(m + p))$$

$$(\exists n > \exists m) \leftrightarrow (\exists p)(\exists n \equiv \exists(m + p)).$$

Multiplication may be quite simply defined for the weak quantifiers in terms of one–many relations, for example

$$1 - m(\Re) \quad \text{for} \quad (\forall \alpha \beta \gamma)(\alpha \Re \gamma \ \& \ \beta \Re \gamma \to \alpha = \beta)$$

$$(\exists(n \cdot m)\alpha)(\Phi\alpha) \quad \text{for} \quad (\exists \Re)(1 - m(\Re) \ \& \ (\exists n\alpha)(\exists m\beta)(\alpha \Re \beta \ \& \ \Phi\beta)).$$

Alternatively we could use the counting relation of being mutually exclusive, defining

$$\Phi \ excl \ \Psi \quad \text{for} \quad \sim(\exists \alpha)(\Phi\alpha \ \& \ \Psi\alpha)$$

$$(\exists(n \cdot m)\alpha)(\Phi\alpha) \quad \text{for} \quad (\exists n_{excl}\psi)((\exists m\alpha)(\psi\alpha) \ \& \ (\forall \alpha)(\psi\alpha \to \Phi\alpha)).$$

From either of these definitions we may establish the recursive properties of multiplication

$$\exists(1 \cdot m) \equiv \exists m$$

$$(\exists n' \cdot m) \equiv \exists(n \cdot m + m)$$

and therefrom the remaining properties. With strong quantifiers the definitions would be more complicated, and indeed the second could not be reproduced without using the special definition of strong quantifiers with counting relations other than non-identity which was canvassed on p. 16. An approach to the first would be

$$((n \cdot m)\alpha)(\Phi\alpha) \quad \text{for} \quad (\exists \Re)((\forall \beta)(\Phi\beta \to (1\alpha)(\alpha \Re \beta))$$

$$\& \ (n\alpha)(\exists \beta)(\alpha \Re \beta)$$

$$\& \ (\forall \alpha)((\exists \beta)(\alpha \Re \beta) \to (m\beta)(\alpha \Re \beta \ \& \ \Phi\beta))).$$

However, this is somewhat unwieldy, and with the strong quantifiers it would be simpler to use a purely arithmetical definition of multiplication.

As for the arithmetical definitions, there are several well-known methods of exchanging any pair of recursive equations for an explicit definition making use of higher-order quantification. For example the recursive equations

$$1 + m = m'$$
$$n' + m = (n + m)'$$

can be rephrased as

$$1 + m = m'$$
$$(\forall kl)(k + m = l \rightarrow k' + m = l')$$

and these can then be regarded as furnishing information on the relation $\ldots + m = \ldots$, *viz.* that it satisfies the condition on $\Re$

$$1 \Re m' \ \& \ (\forall kl)(k \Re l \rightarrow k' \Re l').$$

Furthermore it is quite easily argued (by induction) that $\ldots + m = \ldots$ is the *minimum* relation that satisfies these conditions, i.e. that

$$(\forall \Re)(1 \Re m' \ \& \ (\forall kl)(k \Re l \rightarrow k' \Re l') \rightarrow (\forall kl)(k + m = l \rightarrow k \Re l)).$$

As with the definition of the ancestral (Vol. 1, Chap. 5, §3), we may therefore define

$$n + m = p \quad \text{for} \quad (\forall \Re)(1 \Re m' \ \& \ (\forall kl)(k \Re l \rightarrow k' \Re l') \rightarrow n \Re p).$$

The recursive equations can then be recovered from this definition, together with the needed information that $\ldots + m = \ldots$ is a many–one relation, and therefore appropriately symbolized with a sign of equality, i.e.

$$n + m = p \ \& \ n + m = q \rightarrow p = q.$$

Essentially the same treatment may be applied to any numerical operation which can be characterized by a pair of recursive equations, and in the presence of Peano's postulates there is no difficulty in establishing the adequacy of the higher-order definition. For example multiplication would by this method be defined

$$n \cdot m = p \quad \text{for} \quad (\forall \Re)(1 \Re m \ \& \ (\forall kl)(k \Re l \rightarrow k' \Re (l + p)) \rightarrow n \Re p).$$

One point worth noticing is that this method is not available without further discussion unless (in the present case) numbers are actually identified with numerical quantifiers. This is because the method requires us to use a variable for arbitrary (dyadic) predicates of numbers, and although we are already employing variables for predicates of quantifiers this is of no assistance if numbers are distinguished from quantifiers. For example if we introduce arithmetical statements by defining, as in Volume 1 (Chap. 5, §4),

$$n =_N m \quad \text{for} \quad (\forall \Re)(\exists n_\Re \equiv \exists m_\Re)$$

$$n \mathbf{S}_N m \quad \text{for} \quad (\forall \Re)(\exists n_\Re \equiv \exists m'_\Re)$$

then these two formulae are introduced *as wholes*, and the variable n does not actually take the place of any quantifier expression when it occurs in an arithmetical context. So far as arithmetical statements are concerned, then, we have so far introduced numerical variables only in two very special contexts, and it cannot be assumed without more ado that Φn or $n \Re m$ are intelligible expressions. I shall have more to say on this problem later (pp. 249 *et seq.*), but while we are still concerned with the elementary arithmetic of natural numbers it seems better just to avoid it.

This we can do by employing an alternative method of 'purely arithmetical' definition, which in effect relies on the idea of *powers* of a relation. Intuitively these are easily understood from the recursive equivalences[8]

$$\alpha \Re^1 \beta \leftrightarrow \alpha \Re \beta$$

$$\alpha \Re^{n'} \beta \leftrightarrow (\exists \gamma)(\alpha \Re^n \gamma \ \& \ \gamma \Re \beta).$$

Supposing for the moment that these suffice to introduce the notion, we may then define addition in terms of powers of the relation $\mathbf{S}_N$ by putting

$$n + m =_N p \quad \text{for} \quad p(\mathbf{S}_N)^m n$$

and we may define multiplication in terms of powers of the relation $(\mathbf{S}_N)^m$ by putting[9]

$$n \cdot m =_N p \quad \text{for} \quad (n =_N 1 \ \& \ m =_N p) \lor (\exists q)(n \mathbf{S}_N q \ \& \ p((\mathbf{S}_N)^m)^q m)$$

---

[8] If the 0th power is to be included, the first clause is usually given as
$$\alpha \Re^0 \beta \leftrightarrow \alpha = \beta$$
(but see p. 26 for an alternative suggestion).

[9] Or more simply, if 0 is included
$$n \cdot m =_N p \quad \text{for} \quad p((\mathbf{S}_N)^n)^m 0.$$

Now in the special case of the relation $\mathbf{S}_N$ we can in fact define its powers in terms of the strong numerical quantifiers, by putting[10]

$$p(\mathbf{S}_N)^m n \quad \text{for} \quad (m_{\neq N} q)(p*\mathbf{S}_N q \ \& \ q \divideontimes \mathbf{S}_N n)$$

and the same definition is also available for powers of the relation $(\mathbf{S}_N)^m$. So this is another way of giving explicit and 'purely arithmetical' definitions of addition and multiplication. But of course our ability to define the powers of the relation $\mathbf{S}_N$ in this way depends on some very special features of that relation, and it would be natural to want a much more general definition of powers of relations. This brings me to my next section.

2. Finite Powers

Whereas the numerical quantifiers evidently reflect our usage of the numerals 'one', 'two', 'three', . . ., I now wish to consider some numerical concepts which are naturally expressed by different series of numerical expressions. In the following section I shall consider the ordinal adjectives 'first', 'second', 'third', . . ., and in the present section the numerical adverbs 'once', 'twice', 'three times', ...

These adverbs have a variety of uses. One important use is in contexts of the form 'n times as . . . as' (e.g. 'twice as tall as', 'three times as heavy as'), and this use is the main topic of the next chapter. Another is their use to mean 'on n occasions', as in 'I fell over twice', and this use is easily analysed in terms of numerical quantifiers applied to occasions or times (*viz.* 'there are two times t such that I fell over at t'). However, there is also a somewhat different use which is no doubt what Wittgenstein had in mind when he claimed in *Tractatus* 6.021 that 'a number is the exponent of an operation'. For example if a child is told to 'add 1 to 2 three times' he is not likely to be given full marks if he writes down the sentence '$2 + 1 = 3$' three times. Rather, he is expected to write this sentence only once, and after that to write '$3 + 1 = 4$', and after that '$4 + 1 = 5$', or alternatively to

---

[10] Of if 0 is included

$$p(\mathbf{S}_N)^m n \quad \text{for} \quad p*\mathbf{S}_N n \ \& \ (m_{\neq N} q)(p*\mathbf{S}_N q \ \& \ q \divideontimes \mathbf{S}_N n).$$

write just '$2 + 1 + 1 + 1 = 5$'. The operation of adding 1 is indeed to be performed three times (on three occasions), but what the operation is to be performed on is different on each occasion: the result of performing it once is to be taken as the starting point for performing it again, and so on. It is this use which I wish to consider in the present section.

This use is easily explained in terms of the numerical powers of relations, for if we write xRy to signify that x is what results from performing the operation on y, then performing the operation on y *n times* (in this sense) is the same as finding the thing x such that xR^ny. So, in our example, the child is being told to find the number n such that $n(\mathbf{S}_N)^32$. Similarly if $n' \cdot m$ is explained as the number which results upon adding m to itself n times we have just the explanation given on p. 22 using powers of the relation $(\mathbf{S}_N)^m$. Now we make use of the idea of applying an operation several times (in this sense) only when the operation is a unary one and so can be represented by a binary one–many relation. However, numerical powers are usually defined for all binary relations, whether or not they are one–many, and so it is clear that the latter notion is more general. It is probably true that most of the *useful* applications of numerical powers of relations do concern one–many relations where we could always regard the numerical index as the exponent of an operation if we wished to. It also seems to be true that English has no very natural way of expressing powers of relations except as exponents of operations, and even here the locution is not altogether common. For example we certainly would not characterize a grandfather as one who was a father 'twice over', and it would be unusual to be told that a grandfather is one who results from twice applying the operation of taking the father of. However, this seems to be a point of idiom which has no very noticeable significance, and since the idea of numerical powers of relations is in fact both more general and perfectly clear I shall henceforth treat of this rather than of numerical exponents of operations. But the discussion could easily be specialized to operations if that was thought more appropriate.

Now when Wittgenstein claimed that a number *is* the exponent of an operation, he presumably meant at least to imply that numbers can be introduced in this role without relying on any antecedent understanding of numbers in any other role, but since he gave only the merest sketch in the *Tractatus* of how such an introduction might proceed, his claim has largely been ignored by subsequent philosophers (including, of course, the later Wittgenstein). However, the claim is in fact entirely correct, and the main purpose of this section is to demonstrate that fact by outlining an appropriate construction of arithmetic based on this application of numbers and presupposing nothing else about numbers (except that, as I say, I shall consider relations in general and not just operations). What makes the construction possible is, once more, that numbers are *type-neutral* in this role, for without the type-neutrality a proof of Peano's postulates would have to rely upon an axiom of infinity.

We begin, then, by introducing the general idea of an operation on a dyadic relation which yields another dyadic relation of the same type and which is itself a type-neutral operation in that it can be applied to *all* dyadic relations, whatever their type, provided only that they are *homogeneous* relations, i.e. that their terms are of the same type as each other. Let us assume that $\Re$, $\mathfrak{S}$, $\mathfrak{T}$, ... are already introduced as schematic letters for homogeneous dyadic relations of any type, so that our problem is to introduce a new series of letters, say i, j, ..., to represent the type-neutral operators on these relations. We shall write $\Re^i$ as an expression for the relation which results by applying the operator i to the relation $\Re$, and in outline the appropriate explanation will be that a proposition is of the form $\alpha\Re^i\beta$ if an only if (i) it is also of the form $\alpha\mathfrak{S}\beta$ and (ii) it may be analysed as containing a component relation $\Re$ and as constructed from that relation in a type-neutral way, e.g. by quantification and truth-functional combination. Construction from the same component relation will be signified by repetition of the same $\Re$-letter, as in $\Re^i$ and $\Re^j$, while construction in the same type-neutral way will be signified by repetition of the same i-letter, as in $\Re^i$ and $\mathfrak{S}^i$. (Familiar examples of relations of the form $\Re^i$ are, for example, $\bar{\Re}$, $\breve{\Re}$, $*\Re$.) This introduction of i-letters and $\Re$-letters may correspond fairly closely to my earlier introduction of type-neutral Q-letters and Φ-letters, though the details will be more complicated since $\Re$-letters are dyadic while Φ-letters are monadic. Anyway I shall say no more about this feature of the underlying logic.

Supposing that the basic notation is introduced, we may now

turn to the fundamental definitions, and here we have first to decide whether 0 should be included as a relational power. Nothing of importance in fact hangs on the decision, but if 0 is to be included I would propose a small modification to the usual definition. Normally the 0th power of any relation is taken to be identity, and this identity is in turn taken as obeying the Leibnizian definition. However, we have seen that in any construction of the natural numbers we shall want to introduce a special identity relation between numbers which cannot be assumed to be Leibnizian without an axiom to the purpose. Besides it is often useful to be able to take a more inclusive relation as the 0th power of a relation, quite apart from any special considerations about numerical identity. For example if $\Re$ is the relation between classes of having one more member than, then $\Re^n$ will be the relation of having n more members than in every case except where $n = 0$, and in the latter case it will be identity according to the usual definition. Clearly it will be much more convenient for many purposes to take $\Re^0$ as a relation in the same family as all other powers of $\Re$, and a reasonable suggestion with the present example would be to identify it with the relation of having the same number of members as. What I would suggest, then, is that instead of saying that $\alpha \Re^0 \beta$ is to hold if and only if α and β are *absolutely* indistinguishable, a more useful general condition will be that it is to hold if and only if α and β are indistinguishable, *so far as the relation $\Re$ is concerned*. This will lead us to the definition

$$\alpha \Re^0 \beta \quad \text{for} \quad (\forall \gamma)(\alpha \Re \gamma \leftrightarrow \beta \Re \gamma) \, \& \, (\forall \gamma)(\gamma \Re \alpha \leftrightarrow \gamma \Re \beta)$$

which evidently secures the desired result for the present example. To this definition we add, quite generally

$$\alpha \Re^{i'} \beta \quad \text{for} \quad (\exists \gamma)(\alpha \Re^i \gamma \, \& \, \gamma \Re \beta)$$

though of course we shall only be interested in the relation $\Re^{i'}$ where 'i' is a numeral. We may note, incidentally, that from these definitions we can deduce the usual equivalence

$$\alpha \Re^{0'} \beta \leftrightarrow \alpha \Re \beta$$

which shows that my unusual suggestion about $\Re^0$ does not affect the interpretation of any other power of $\Re$.

Next we may turn to the task of singling out the operations of raising to a numerical power from all other type-neutral operations on homogeneous dyadic relations. We may begin by noting three convenient abbreviations

$$\alpha \Re | \mathfrak{S} \beta \quad \text{for} \quad (\exists \gamma)(\alpha \Re \gamma \ \& \ \gamma \mathfrak{S} \beta)$$

$$\Re \subseteq \mathfrak{S} \quad \text{for} \quad (\forall \alpha \beta)(\alpha \Re \beta \to \alpha \mathfrak{S} \beta)$$

$$\Re \equiv \mathfrak{S} \quad \text{for} \quad (\forall \alpha \beta)(\alpha \Re \beta \leftrightarrow \alpha \mathfrak{S} \beta)$$

and using the last of these we may offer definitions of identity and succession for our operators, *viz.*

$$i = j \quad \text{for} \quad (\forall \Re)(\Re^i \equiv \Re^j)$$

$$i \, \mathbf{S} j \quad \text{for} \quad (\forall \Re)(\Re^i \equiv \Re^{j'}).$$

(Clearly $i \mathbf{S} j$ is a variant on $i = j'$.) The identity here defined is not assumed to be Leibnizian, but it does of course have some substitutive properties, and in particular

$$i = j \to (\forall k)(i \mathbf{S} k \leftrightarrow j \mathbf{S} k)$$

$$i = j \to (\forall k)(k \mathbf{S} i \leftrightarrow k \mathbf{S} j).$$

It follows that this identity will licence substitution in all contexts built up from $\mathbf{S}$ and $=$. We may also note that since the operation of taking the 0th power and of taking the i'^{th} power are clearly type-neutral (wherever taking the ith power is type-neutral) we have by the substitution rules for i-letters

$$(\exists i)(i = 0)$$

$$(\exists i)(i = j').$$

The next step will be to introduce the proper ancestral $*\mathbf{S}$ of the relation $\mathbf{S}$, and unless we suppose that we already have general axioms available for the ancestral of a relation this will have to be done by explicit definition, and this in turn will require us to use a variable for predicates of our type-neutral operators. However, there is no need for any new symbol at this point, since our letter $\Re$ is supposed to stand for homogeneous dyadic relations of *any* type and therefore *inter alia* for the type we require. We therefore define

$$i *\mathbf{S} j \quad \text{for} \quad (\forall \Re)((\mathbf{S} \subseteq \Re) \ \& \ (\mathbf{S} | \Re \subseteq \Re) \to i \Re j)$$

(where the variables concealed in the antecedent clauses are of course taken to be i-variables). We then introduce a definition of the improper ancestral

$$i*\mathbf{S}j \quad \text{for} \quad i=j \vee i*\mathbf{S}j$$

which is marginally non-standard because it uses our defined identity and not Leibnizian identity. Finally we define the operations of raising to a numerical power as those type-neutral operations which bear $*\mathbf{S}$ to 0, so that

$$(\forall n)(-n-)$$

$$(\exists n)(-n-)$$

abbreviate (respectively)

$$(\forall i)(i*\mathbf{S}0 \to (-i-))$$

$$(\exists i)(i*\mathbf{S}0 \mathbin{\&} (-i-)).$$

From the properties of the ancestral we easily deduce first

$$(\exists n)(n=0)$$

$$(\exists n)(n=m')$$

and second that mathematical induction holds for our numerical variables. We also have as immediate consequences of our definitions[11]

$$0 \neq n'$$

$$n=m \to n'=m'.$$

As before, then, our problem is to establish the converse of this last thesis.

As a preliminary we may note that we have as a trivial consequence of induction

(1) $n=0 \vee (\exists m)(n=m')$

Next we introduce two auxiliary definitions. First

$$\mathfrak{R}^{n+m} \quad \text{for} \quad \mathfrak{R}^{n} \mid \mathfrak{R}^{m}.$$

---

[11] To see that $0 \neq n'$, notice that if $\mathfrak{R}$ is any contradictory relation it will be true that $(\forall \alpha)(\alpha \mathfrak{R}^{0} \alpha \mathbin{\&} \sim \alpha \mathfrak{R}^{n'} \alpha)$.

From this definition we can immediately deduce the usual recursive equations for addition

(2) $n + 0 = n$

$\quad n + m' = (n + m)'$

for the first is obtained simply by applying the definition of $\Re^0$ and the second is a consequence of the perfectly general theorem

$$(\Re \mid \mathfrak{S}) \mathfrak{T} = \Re \mid (\mathfrak{S} \mid \mathfrak{T}).$$

From these recursive equations we can deduce by the standard method such properties of addition as[12]

$$(\exists p)(n + m = p)$$

$$n + m = m + n$$

$$n + (m + p) = (n + m) + p$$

$$n = m \rightarrow (n + p = m + p)$$

and of course

$$n + 0' = n'.$$

I shall cite all these properties of addition henceforth simply as theorem (2). Our second auxiliary definition is

$$n \geqslant m \quad \text{for} \quad (\exists p)(n = m + p)$$

and immediate consequences of this definition are

(3) $n = m \rightarrow n \geqslant m$

(4) $n \geqslant 0$

(5) $n \geqslant m \rightarrow n' \geqslant m'$

(6) $0 \not\geqslant n'.$

By applying theorem (1) and the recursive equations for addition we can also prove quite straightforwardly

(7) $n \geqslant m \ \& \ n \neq m \rightarrow n \geqslant m'$

and hence we further have, by induction,

(8) $n \neq m \rightarrow n \geqslant m' \lor m \geqslant n'.$

---

[12] It is in fact often simpler to prove these theorems directly from the definition of $+$ than it is to work only from the recursive equations.

(The induction basis follows from theorems (4) and (5), and the induction step from theorems (3), (5), and (7).) With so much established by way of preliminary, we may now turn to the crucial argument of this construction.

Consider first what is needed to establish the very simple thesis $1 \neq 2$. According to our present definitions $1 = 2$ if and only if $(\forall \Re)(\Re^1 \equiv \Re^2)$, and it is easily seen that this would indeed be true if there were at most one object and $\Re$ were confined to first-level relations holding only between objects. However, as it is $\Re$ is intended as a variable for homogeneous relations of *any* type, so what we have to do is just to find a homogeneous relation of some type or other that has two distinct terms in its field. We are already in possession of such a relation, namely the successor relation between powers of relations, for $1S0$ is a matter of definition and $1 \neq 0$ is something we have just established. This is sufficient for our purpose, for consider the relation $\mathbf{T}$ so defined that

$$(\forall nm)(n\mathbf{T}m \to n\mathbf{S}m \ \& \ m \neq 1).$$

From what we have just said it follows that

$$1\mathbf{T}^10$$

and it is also clear from the definition of $\mathbf{T}$ that

$$\sim 1\mathbf{T}^20$$

for

$$1\mathbf{T}^20 \to (\exists n)(1\mathbf{T}n \ \& \ n\mathbf{T}0)$$

$$\to (\exists n)(n \neq 1 \ \& \ n\mathbf{S}0)$$

$$\to (\exists n)(n \neq 1 \ \& \ n = 1).$$

Hence $1 \neq 2$ follows, and in fact since the same argument suffices to prove $\sim 1\mathbf{T}^{n+2}0$, we may conclude more generally that $1 \neq n + 2$. However, it is easily seen that we cannot stop at just two terms, for we shall need a relation of three distinct terms to disprove the thesis $2 = 4$, a relation of four distinct terms to disprove the thesis $3 = 6$, and so on. Since in fact the relation $\mathbf{S}$ has infinitely many distinct terms, this is what we shall now prove.

The general strategy of the argument can be described thus.

What we have just done is to use the premise that $0 \neq 1$ to establish further that $1 \neq 2$, and we have already established that $0 \neq 2$. Hence the argument in effect moves from the existence of two distinct terms of the relation $\mathbf{S}$, *viz.* 0 and 1, to the existence of three distinct terms, *viz.* 0, 1, and 2. We can now use the same pattern of argument to prove the existence of four distinct terms, *viz.* 0, 1, 2, and 3. From what we have already we know that $0 \neq 3$ and that $1 \neq 3$, and by considering the new relation $\mathbf{T}$ so defined that

$$(\forall nm)(n\mathbf{T}m \leftrightarrow n\mathbf{S}m \; \& \; m \neq 2)$$

we shall again show that

$$2\mathbf{T}^2 0 \; \& \; \sim 2\mathbf{T}^3 0$$

whence $2 \neq 3$. With four distinct terms available we may then show that 4 is not to be identified with any of them, and hence that we have five distinct terms, and so on. Formulating this pattern of argument as a general inductive argument (which I shall do in a moment) we shall thus prove the existence of infinitely many distinct terms to the relation $\mathbf{S}$, and the deduction of the missing postulate will then be quite straightforward. One might say that in this argument we prove the existence of infinitely many 'numbers', i.e. powers of relations, by relying on the fact that each 'number' n bears the nth power of the successor relation, and no other power, to the 'number' 0. This is a self-application of the idea of numerical powers of relations closely similar to the previous self-application whereby we proved the existence of infinitely many 'numbers', i.e. numerical quantifiers, by relying on the fact that for each 'number' n the numerical quantifier 'there are n. . .', and no other numerical quantifiers, applies to the 'numbers' less than n.

I now proceed to the general inductive argument. The chief lemma is

(9) $n\mathbf{S}^m 0 \leftrightarrow n = m$

and this may be proved quite straightforwardly by induction on m. For $m = 0$ what has to be proved is (in view of the definition of $\Re^0$) that

$$((\forall p)(n = p' \leftrightarrow 0 = p') \; \& \; (\forall p)(p = n' \leftrightarrow p = 0')) \leftrightarrow n = 0.$$

The implication from right to left is immediate from the definition of $=$, and the implication from left to right results from the observation that the first clause of the left-hand side implies

$$(\exists p)(n = p') \to (\exists p)(0 = p')$$

whence by theorems (3) and (6) we have

$$\sim(\exists p)(n = p')$$

and so by theorem (1)

$$n = 0.$$

This establishes the induction basis[13]

$$n\mathbf{S}^0 0 \leftrightarrow n = 0$$

The induction step is then utterly straightforward in view of the fact that (by theorem (2))

$$n\mathbf{S}^{m'} 0 \leftrightarrow (\exists p)(n\mathbf{S}p \ \& \ p\mathbf{S}^m 0).$$

This establishes our chief lemma.

Now we shall organize the argument as a proof by induction of $n \not\geq n'$. The induction basis is already given by theorem (6), so we next assume as hypothesis of induction

(i) $n \not\geq n'$.

Now suppose (for *reductio ad absurdum*) that

$$n\mathbf{S}^m n'.$$

Then by theorem (9) we have

$$n\mathbf{S}^m n' \ \& \ n'\mathbf{S}^{n'} 0$$

whence by the definition of addition and theorem (2)

$$n\mathbf{S}^{n' + m} 0$$

whence by theorem (9) again

$$n = n' + m$$

---

[13] Note that with $\Re^0$ defined as *Leibnizian* identity this theorem would fail, because of the two distinct notions of identity involved, and this would introduce a tedious complication to the proof. With the present definition we of course have quite generally

$$n\mathbf{S}^0 m \leftrightarrow n = m.$$

which, by the definition of $\geqslant$, contradicts (i). Hence our inductive hypothesis implies

(ii) $(\forall m)(\sim n\mathbf{S}^m n')$.

We now introduce a relation $\mathbf{T}$ so defined that

$$(\forall pq)(p\mathbf{T}q \leftrightarrow p\mathbf{S}q \,\&\, q \neq n').$$

From this definition it is perfectly simple to show by induction on m that

(iii) $(\forall pq)(p\mathbf{T}^{m'}q \rightarrow p\mathbf{S}^{m'}q \,\&\, q \neq n')$.

By induction on m again we may also show that

(iv) $(\forall p)(n\mathbf{S}^m p \rightarrow n\mathbf{T}^m p)$.

It is easily checked that (iv) holds for $m = 0$, so, taking (iv) now as hypothesis of induction, assume

$$n\mathbf{S}^{m'}p.$$

Then in view of (ii) we have

$$n\mathbf{S}^{m'}p \,\&\, p \neq n'$$

which is to say

$$(\exists q)(n\mathbf{S}^m q \,\&\, q\mathbf{S}p \,\&\, p \neq n').$$

By the definition of $\mathbf{T}$ we therefore have

$$(\exists q)(n\mathbf{S}^m q \,\&\, q\mathbf{T}p)$$

and hence by the inductive hypothesis

$$(\exists q)(n\mathbf{T}^m q \,\&\, q\mathbf{T}p)$$

and so finally

$$n\mathbf{T}^{m'}p$$

as required. This establishes (iv).

Now let us pull the pieces together. In view of (i) we have

$$n'\mathbf{T}n$$

and in view of (iv) we have, using theorem (9),

$$n\mathbf{T}^n 0$$

whence on the one hand

(v) $n'\mathbf{T}^{n'}0$.

But also if

$$n'\mathbf{T}^{n'+m'}0$$

then by the definition of addition and theorem (2) we have

$$(\exists q)(n'\mathbf{T}^{m'}q \ \& \ q\mathbf{T}^{n'}0)$$

and hence, using (iii),

$$(\exists q)(q \neq n' \ \& \ q\mathbf{S}^{n'}0)$$

which contradicts theorem (9). So this shows that on the other hand we have

(vi) $\sim n'\mathbf{T}^{n'+m'}0$.

Bringing (v) and (vi) together, and applying the definition of $=$, we have

$$n' \neq n' + m'$$

whence by theorem (2)

$$n' \neq n'' + m$$

which is the desired conclusion

$$n' \not\geqslant n''.$$

This completes the argument, and we have established the crucial theorem

(10) $n \not\geqslant n'$.

The last of Peano's postulates is now at once forthcoming, for we evidently have

$$n' = m' \ \& \ (n \geqslant m' \lor m \geqslant n') \to (n \geqslant n' \lor m \geqslant m')$$

so our theorem (10) implies

$$n' = m' \to \sim(n \geqslant m' \lor m \geqslant n')$$

and hence by theorem (8) we deduce

$$n' = m' \to n = m$$

as desired. This completes the construction.

Some details of the underlying logic may require a little more filling out[14], but it should be clear that the crucial feature of the construction is its use of the concept of type-neutrality, whereby the thesis

$$n'\mathbf{T}^{n'}0 \ \& \ \sim n'\mathbf{T}^{n'+m'}0$$

is taken as a counter-example to the generalization

$$(\forall \Re)(\forall \alpha \beta)(\alpha \Re^{n'} \beta \leftrightarrow \alpha \Re^{n'+m'} \beta).$$

The relevant concept of type-neutrality is very similar to the concept used earlier in constructions based on the numerical quantifiers, and a defence of the one could, I think, always be adapted to a defence of the other. Therefore, construed as a logicist construction of the theory of natural numbers, the present example is surely both distinct from, and no worse than, the previous ones.

Indeed it might be argued that it is better. We have already noticed that on this approach to the natural numbers we may give a particularly simple definition of addition, *viz.*

$$\Re^{n+m} \quad \text{for} \quad \Re^n | \Re^m$$

and we could also add a particularly simple definition of multiplication

$$\Re^{n \cdot m} \quad \text{for} \quad (\Re^n)^m.$$

Alternatively, if addition and multiplication are to be introduced as concepts of 'pure arithmetic' (disregarding their applications), then again an introduction using the powers of the successor relation is particularly simple and corresponds to quite a natural way of explaining these operations (e.g. '$n + 3$ is the

---

[14] For example, are we justified in assuming the existence of 'a relation $\mathbf{T}$ so defined that $(\forall pq)(p\mathbf{T}q \leftrightarrow p\mathbf{S}q \ \& \ q \neq n')$'? To eliminate the introduction of a special *symbol* we may of course rephrase the inductive step so that it contains a proof of

$$(\forall \Re)((\forall pq)(p\Re q \leftrightarrow p\mathbf{S}q \ \& \ q \neq n') \to (n'\Re^n 0 \ \& \ \sim n'\Re^{n'+m'}0))$$

whence by ordinary quantifier rules we have

$$(\exists \Re)((\forall pq)(p\Re q \leftrightarrow p\mathbf{S}q \ \& \ q \neq n') \to (\exists \Re)(n'\Re^n 0 \ \& \ \sim n'\Re^{n'+m'}0)$$

The licence to assert the antecedent is then given by the substitution rules for predicates ('remainders') of type-neutral operators on homogeneous dyadic relations (cf. Vol. 1, pp. 69–72).

number 3 steps on from n'). In these respects, then, the construction using powers of relations must be admitted to have a certain elegance. What is more interesting is that numerical powers could be argued to be more general than numerical quantifiers, for we can define the quantifiers in terms of powers by using the equivalence

$$(\exists n_{\Re}'\alpha)(\Phi\alpha) \leftrightarrow (\exists \ominus)((\exists\alpha\beta)(\alpha\ominus^n\beta \ \& \ \Phi\beta) \ \& \ (\forall\alpha\beta)(\alpha\ominus\beta \ \& \ \Phi\beta\rightarrow\Phi\alpha)$$
$$\& \ (\forall\alpha\beta)(\alpha*\ominus\beta\rightarrow\alpha\Re\beta)).$$

On the other hand a definition of numerical powers in terms of numerical quantifiers must be altogether more roundabout (p. 53).

In view of these facts it does seem rather surprising that the English idiom provides no natural way of expressing the numerical powers of relations except in a few rather special cases. Perhaps it might be suggested that what could be expressed by using powers of relations is actually expressed in English by using the *ordinal* numerals 'first', 'second', 'third', and so on. Indeed it may be noticed that I have myself had to use ordinal numerals in this section in order to characterize particular numerical powers (*viz.* as *the nth power* of the relation in question). The topic of ordinal numerals brings me to my next section.

3. Finite Ordinals

The numerical adjectives 'first', 'second', 'third', . . . are called *ordinal* numerals evidently because their application to things depends on the *order* in which those things are considered. Now it is usual to think of an order as generated by an *ordering relation*, by which is meant any relation that is transitive, asymmetrical, and connected, at least among the things that we are currently considering. That is, the theses

(i) $\alpha\Re\beta \ \& \ \beta\Re\gamma \rightarrow \alpha\Re\gamma$

(ii) $\sim(\alpha\Re\beta \ \& \ \beta\Re\alpha)$

(iii) $\alpha \neq \beta \rightarrow (\alpha\Re\beta \lor \beta\Re\alpha)$

should hold for the relation $\Re$, at least when the variables are confined in the appropriate way. However, it soon becomes clear that our ordinary idiom is by no means so exacting in its requirements on the relations that can be used in connection with ordinals. For example there is nothing evidently wrong with the locution 'the third man from the left', which clearly invokes the relation 'to the left of' to supply the relevant order, but this relation is certainly not in general connected. One might suggest that any sensible use of this phrase will require that some particular collection of men is singled out as the one we have in mind, in fact a line of men, such that the relation 'to the left of' *is* connected when confined to the members of that collection. This would be too neat, however. The collection we have in mind may be for example all the men in a certain room, and they may be scattered about the room in any way at all provided that *the three leftmost* men form a (rough) line so that of any two of them one is to the left of the other. I would suggest, then, that it would be more faithful to our ordinary idiom if the requirement of connexity is introduced in another way.

The relations that are used in conjunction with ordinals are surely transitive and asymmetrical[15], but in place of connexity I propose to substitute the weaker requirement that results by replacing the condition of non-identity in (iii) by the condition of distinguishability under $\Re$, i.e. any two of the items we are concerned with must be such that either one of them bears $\Re$ to the other, or they each bear $\Re$ to the same things and the same things bear $\Re$ to each of them. A conveniently short way of formulating this condition is

(iii′) $\alpha\Re\beta \rightarrow (\gamma\Re\beta \vee \alpha\Re\gamma)$

A relation $\Re$ that satisfies (i), (ii), and (iii′) when the variables are confined to some collection will be said to be a *weak ordering* of that collection, so I define[16] (omitting the qualification 'weak' for brevity)

---

[15] See note 16.

[16] It is often convenient to take $\leqslant$ rather than $<$ as a paradigm ordering relation as on pp. 73–6. The appropriate 'strong' conditions are then

(i) $\alpha\Re\beta \ \& \ \beta\Re\gamma \rightarrow \alpha\Re\gamma$

(ii) $\alpha\Re\beta \ \& \ \beta\Re\alpha \rightarrow \alpha = \beta$

(iii) $\alpha\Re\beta \vee \beta\Re\alpha.$

The analogous 'weak' conditions are obtained simply by omitting (ii) and leaving (i) and (iii) unchanged (cf. p. 101).

$$Ord(\Re, \Phi)$$

for

$$(\forall\alpha\beta\gamma)(\Phi\alpha \ \& \ \Phi\beta \ \& \ \Phi\gamma \to (\alpha\Re\beta \ \& \ \beta\Re\gamma \to \alpha\Re\gamma))$$
$$\& \ (\forall\alpha\beta)(\Phi\alpha \ \& \ \Phi\beta \to \sim(\alpha\Re\beta \ \& \ \beta\Re\alpha))$$
$$\& \ (\forall\alpha\beta\gamma)(\Phi\alpha \ \& \ \Phi\beta \ \& \ \Phi\gamma \to (\alpha\Re\beta \to (\gamma\Re\beta \lor \alpha\Re\gamma))).$$

We may also add that a relation is, absolutely, a weak ordering
if it is a weak ordering of its total field:

$$Ord(\Re) \quad \text{for} \quad (\forall\phi)((\forall\alpha)(\phi\alpha \leftrightarrow (\exists\beta)(\alpha\Re\beta \lor \beta\Re\alpha)) \to Ord(\Re, \phi)).$$

Now let us return to ordinals.

Where ordinals occur in conjunction with ordering relations I
think that all occurrences can be viewed as having the general
form 'α is the nth Φ under $\Re$'—more briefly 'n^{th} $(\alpha, \Phi, \Re)$'—
where it will be presupposed at least that $\Re$ weakly orders the
Φ's. In the normal case it will also be implied that $\Re$ strongly
orders the Φ's at least up to and including α. Now this will be
implied if we adopt the strong recursive equivalences

A. $\begin{cases} 1^{\text{st}}(\alpha, \Phi, \Re) \leftrightarrow \Phi\alpha \ \& \ (\forall\gamma)(\Phi\gamma \to \alpha = \gamma \lor \alpha\Re\gamma) \\ n'^{\text{th}}(\alpha, \Phi, \Re) \to (\exists\beta)(n^{\text{th}}(\beta, \Phi, \Re) \ \& \ \Phi\alpha \ \& \ \beta\Re\alpha \ \& \\ \qquad\qquad\qquad\qquad (\forall\gamma)(\Phi\gamma \ \& \ \beta\Re\gamma \to \alpha = \gamma \lor \alpha\Re\gamma)). \end{cases}$

An alternative procedure would be to adopt the weak recursive
equivalences

B. $\begin{cases} 1^{\text{st}}(\alpha, \Phi, \Re) \leftrightarrow \Phi\alpha \ \& \ (\forall\gamma)(\Phi\gamma \to \sim\gamma\Re\alpha) \\ n'^{\text{th}}(\alpha, \Phi, \Re) \leftrightarrow (\exists\beta)(n^{\text{th}}(\beta, \Phi, \Re) \ \& \ \Phi\alpha \ \& \ \beta\Re\alpha \ \& \\ \qquad\qquad\qquad\qquad (\forall\gamma)(\Phi\gamma \ \& \ \beta\Re\gamma \to \sim\gamma\Re\beta)) \end{cases}$

and we could then state separately the requirement that $\Re$
strongly orders the Φ's at least up to and including α as

C. $(\forall\beta\gamma)(\Phi\beta \ \& \ \Phi\gamma \ \& \ \sim\alpha\Re\beta \ \& \ \sim\alpha\Re\gamma \to (\beta = \gamma \lor \gamma\Re\beta \lor \beta\Re\gamma)).$

Of course it is easily seen that we could dispense with recursive
equivalences altogether by defining our ordinals outright in
terms of strong numerical quantifiers. We state separately the
requirement C and define (including the case $n = 0$)

$$n'^{\text{th}}(\alpha, \Phi, \Re) \quad \text{for} \quad \Phi\alpha \ \& \ (n\beta)(\Phi\beta \ \& \ \beta\Re\alpha).$$

Alternatively we could define ordinals in terms of powers of
relations, for where C holds there will be a one–one relation
going step by step from the first Φ to α, and α will be the n'^{th}
Φ if and only if the first Φ bears the nth power of this relation

to α. However, my present purpose is to show that, since recursive equivalences are available, we can if we wish introduce these ordinals directly and prove that Peano's postulates hold for them, without presupposing numbers in any other role.

We shall once more be faced with a complication over the part that Leibnizian identity plays in our fundamental definitions. This is because the crucial theorem will once more be obtained by applying ordinals to themselves and proving that the ordinal 'n^{th}' is the nth ordinal in the series of ordinals. But as in the earlier constructions identity between ordinals will be introduced by explicit definition, and we cannot assume without more ado that the defined identity will coincide with Leibnizian identity. That means that we must somehow provide for a version of ordinals which does not directly rely on Leibnizian identity, and here we have a choice between two procedures.

One is to take the route adopted in our treatment of numerical quantifiers, where we *first* introduced quantifiers explicitly based on Leibnizian non-identity and *then* introduced the more general notion of a quantifier based on any counting relation. Thus in the present case we should use the strong recursive equivalences A in defining the ordinals '1st' and 'n'^{th}' and then introduce some secondary ordinals, say '$n^{\text{th}}_{\approx}$', for use with equivalence relations $\approx$ that could be taken as non-Leibnizian identity relations. For example we might define, for any relation in place of $\approx$,

$$n^{\text{th}}_{\approx}(\alpha, \Phi, \Re)$$

for

$$(\exists\psi)(n^{\text{th}}(\alpha, \psi, \Re) \ \& \ (\forall\alpha)(\psi\alpha\rightarrow\Phi\alpha)$$
$$\& \ (\forall\alpha)(\Phi\alpha\rightarrow(\exists\beta)(\alpha\approx\beta \ \& \ \psi\beta))).$$

Where $\Re$ weakly orders the Φ's and where $\approx$ is an equivalence relation that at least implies indistinguishability under $\Re$, this would come to the same thing as adopting for our ordinals '$n^{\text{th}}_{\approx}$' the recursive equations A with $=$ replaced throughout by $\approx$. We would then show that the ordinal 'n^{th}' is the $n^{\text{th}}_{\approx}$ ordinal in the series of ordinals for our defined identity relation $\approx$ between ordinals.

The second method, which is simpler in practice, is to take the route adopted with powers of relations and avoid all use of

Leibnizian identity in our fundamental definitions. In this case we adopt the weak recursive equivalences B[17], which we could perhaps regard as characterizing a notion not in fact used in English but quite naturally expressed as 'α is *an* n^{th} Φ under $\Re$'. We then regard the ordinary notion 'α is *the* n^{th} Φ under $\Re$' as arising from this by the addition of condition C for an identity relation appropriate to the type of thing in question. However, the main argument will now be directed to establishing the properties of the ordinal '*an* n^{th}' and showing in fact that it is an nth in the series of ordinals. For illustration I here sketch the second route, but this time I pass over most of the details of the proof.

In order to obtain the recursive equivalences, the principle of mathematical induction, and our two required existential theses, all as a result of explicit definition, we must again begin by viewing our ordinals as a special case of something more general. What is required here is evidently the general concept of a triadic relation which takes an item of some type as its first argument, a monadic predicate applicable to items of that type as its second argument, and a dyadic homogeneous relation again applicable to items of that type as its third argument. Further, we shall require our triadic predicates to be type-neutral in the sense that they are significantly applicable to *any* such triad of arguments. Let us use $\mathcal{M}, \mathcal{N}, \ldots$, as variables for such predicates. The fundamental definitions are then as usual:

$$1^{\text{st}}(\alpha, \Phi, \Re) \quad \text{for} \quad \Phi\alpha \ \& \ (\forall\gamma)(\Phi\gamma \to \sim\gamma\Re\alpha)$$

$$\mathcal{N}'(\alpha, \Phi, \Re) \quad \text{for} \quad (\exists\beta)(\mathcal{N}(\beta, \Phi, \Re) \ \& \ \Phi\alpha \ \& \ \beta\Re\alpha$$
$$\& \ (\forall\gamma)(\Phi\gamma \ \& \ \beta\Re\gamma \to \sim\gamma\Re\alpha))$$

$$\mathcal{N} \equiv \mathcal{M} \quad \text{for} \quad (\forall\gamma)(\forall\phi)(\forall\Re)(\mathcal{N}(\alpha, \phi, \Re) \leftrightarrow \mathcal{M}(\alpha, \phi, \Re))$$

$$\mathcal{N} \mathbf{S} \mathcal{M} \quad \text{for} \quad (\forall\alpha)(\forall\phi)(\forall\Re)(\mathcal{N}(\alpha, \phi, \Re) \leftrightarrow \mathcal{M}'(\alpha, \phi, \Re)).$$

We introduce the usual non-standard definition of the ancestral of the relation $\mathbf{S}$ and thereby give a contextual definition of numerical variables, so that

$$(\forall n)(\text{---}n^{\text{th}}\text{---})$$

---

[17] Or, equivalently, the equivalences A with identity replaced by indistinguishability under $\Re$.

is understood as abbreviating

$$(\forall \mathcal{N})(\mathcal{N} * \mathbf{S}1^{\text{st}} \rightarrow (-\mathcal{N}-)).$$

It then follows at once that mathematical induction applies to our numerical variables, and in view of the type-neutrality of the definitions of 1st and $\mathcal{N}'$ we also have[18]

$$(\exists n)(n^{\text{th}} \equiv 1^{\text{st}})$$

$$(\exists n)(n^{\text{th}} \equiv m'^{\text{th}}).$$

It is also perfectly simple to show straight from the definitions that

$$1^{\text{st}} \not\equiv n'^{\text{th}}$$

$$n^{\text{th}} \equiv m^{\text{th}} \rightarrow n'^{\text{th}} = m'^{\text{th}}$$

and our problem is again to deduce the converse of this last thesis.

To simplify our formulae let us write $n^{\text{th}}(\alpha, \Re)$ to mean that α is n^{th} in the whole field of $\Re$ under $\Re$, i.e.

$$n^{\text{th}}(\alpha, \Re) \quad \text{for} \quad (\forall \phi)((\forall \alpha)(\phi \alpha \leftrightarrow (\exists \beta)(\alpha \Re \beta \vee \beta \Re \alpha)) \rightarrow n^{\text{th}}(\alpha, \phi, \Re)).$$

We now introduce the auxiliary definition

$$n^{\text{th}} \geqslant m^{\text{th}} \quad \text{for} \quad (\forall \Re)(Ord(\Re) \,\&\, (\exists \alpha)(n^{\text{th}}(\alpha, \Re)) \rightarrow (\exists \alpha)(m^{\text{th}}(\alpha, \Re)))$$

and we begin by proving

$$n^{\text{th}} \geqslant 1^{\text{st}}$$

$$n^{\text{th}} \geqslant m^{\text{th}} \rightarrow n'^{\text{th}} \geqslant m'^{\text{th}}.$$

The argument is a little more complicated in this case, but no difficulty of principle arises[19]. It is also convenient to define

$$n^{\text{th}} \approx m^{\text{th}} \quad \text{for} \quad (\forall \Re)(Ord(\Re) \rightarrow (\forall \alpha)(n^{\text{th}}(\alpha, \Re) \leftrightarrow m^{\text{th}}(\alpha, \Re)))$$

and we may then prove

$$n^{\text{th}} \geqslant m^{\text{th}} \,\&\, m^{\text{th}} \geqslant n^{\text{th}} \rightarrow n^{\text{th}} \approx m^{\text{th}}.$$

---

[18] I am regarding 1st as idiomatic for 1th, and m'^{th} as idiomatic for $m^{\text{th}'}$.

[19] It is convenient to begin by showing that

$$n^{\text{th}} \geqslant m^{\text{th}} \rightarrow (Ord(\Re) \,\&\, n^{\text{th}}(\alpha, \Re) \rightarrow (\exists \beta)(m^{\text{th}}(\beta, \Re) \,\&\, \sim \alpha \Re \beta)).$$

Another important lemma is

$$Ord(\Re) \,\&\, n^{\text{th}}(\alpha, \Re) \,\&\, m^{\text{th}}(\beta, \Re) \,\&\, \beta \Re \alpha \rightarrow (\exists \gamma)(m'^{\text{th}}(\gamma, \Re)).$$

Next we introduce an ordering relation by putting

$$n^{\text{th}} < m^{\text{th}} \quad \text{for} \quad n^{\text{th}} \nsucceq m^{\text{th}}$$

and it is fairly straightforward to prove the crucial theorem

$$Ord(<) \; \& \; n^{\text{th}}(n^{\text{th}}, \; <).$$

From this theorem there follows

$$n^{\text{th}} \nsucceq n'^{\text{th}}$$

and we can now complete the proof of Peano's postulates using the relation $\approx$ as the relevant version of identity. What remains is to show that this relation coincides with the identity we originally introduced for ordinals, namely equivalence, and this we can now prove as a consequence of the theorem

$$n^{\text{th}} * S m^{\text{th}} \rightarrow n^{\text{th}} \geqslant m^{\text{th}}.$$

Finally we may add the definition of '*the* n^{th}' originally suggested, and prove if we wish that the ordinal n^{th} is *the* n^{th} ordinal in the series of ordinals for this identity relation.

This construction of the system of finite ordinals is closely analogous to our earlier constructions of the 'counting' numbers, the finite cardinals, and the powers of relations. All rely on a self-application depending on the concept of type-neutrality in a very similar way, and on purely logical grounds each seems as admissible as any other. However, this last construction does seem to suffer from a philosophical drawback which I think does not affect the earlier ones, and that is that we are perfectly happy to use ordinal numerals in contexts where, on the face of it, the present account is simply not applicable. The trouble arises where we are concerned with *sequences* which contain *repetitions*. For example there seems nothing wrong with the view that in the sequence

$$\langle a, \, b, \, a, \, b, \, a \rangle$$

a is the first term *and* the third term *and* the fifth term of the sequence, but on our definition of nth this must be impossible. The reason is of course that this sequence is not in our sense an *order*, i.e. it is not generated by any ordering relation applied to its terms, though no doubt it is an order in the intuitive sense of that expression. On the face of it, then, the recursive equivalences for ordinals which we have been relying on must give the wrong results in this case, and matters are not improved if we turn instead to either of our proposed explicit definitions for ordinals. It is clear that we cannot assign ordinals to the terms of the sequence by the method of counting their predecessors in the sequence, for the third, fourth, and fifth terms in this sequence all have exactly the same predecessors, namely a and b and nothing else. Nor does it help to turn to our other suggested definition in terms of powers of one–one relations, for we have only to change the example to the sequence

$$\langle a, b, b, a \rangle$$

and it is clear that there is no one–one relation that takes us always from each term in the sequence to the next. The objection is, then, that ordinals may properly be applied to the terms of a sequence that contains repetitions, and that in these applications the previous analysis cannot be used, because a repetitive sequence will not be generated by an ordering relation.

The most obvious way to bring the previous analysis to bear is to reflect that for any sequence there *is* in fact an ordering relation which can be regarded as generating the sequence, only it is not a relation that relates the *terms* of the sequence to one another but rather one that relates the *subsequences*. Suppose we say that one sequence is an *initial segment* of another if and only if all the terms of the first occur in the same order at the beginning of the second, but not conversely. Then we may consider the initial segments of a given sequence α under the relation 'is an initial segment of', and it is easy to see that this relation will weakly order the initial segments of α[20]. So we may apply our previous analysis to determine the nth initial segment of α under the relation 'is an initial segment of'. The nth term of α may then be defined as the last term of the nth initial segment of α under this relation[21]. We admit, then, that our previous analysis does not apply *directly* to the idea of the nth term of a sequence, though we can give a definition of this as a secondary use of ordinals which relies on the initial analysis. Of course the direction of explanation might have been reversed:

---

[20] If α contains a subsequence which has no first term the order may not be a strong order.

[21] Strictly this definition leaves the last term of α unnumbered. The error is put right in the formal version given below (p. 47).

if we can first introduce the idea of the nth term of a sequence then we can certainly explain 'α is the nth Φ under $\mathfrak{R}$' as (roughly) 'α is the nth term of the sequence of Φ's generated by $\mathfrak{R}$'[22]. However, this raises the question: *can* we first and independently introduce the idea of the nth term of a sequence? This evidently presupposes that we can at least introduce the idea of a sequence, or at any rate of the sort of sequence to which finite ordinals are applicable. Indeed it may well be said that our previous explanation relied on an understanding of the properties of sequences and so itself demands an elucidation of the notion of a sequence, so let us now turn to consider this.

In the orthodox theory of classes the idea of a finite sequence is sometimes introduced in this way. We define

$$\langle a, b \rangle \qquad \text{for} \quad \{\{a\}, \{a, b\}\}$$

$$\langle a, b, c \rangle \qquad \text{for} \quad \langle\langle a, b \rangle, c \rangle$$

$$\langle a, b, c, d \rangle \quad \text{for} \quad \langle\langle a, b, c \rangle, d \rangle$$

and so on. I think it would be generally admitted that this is a highly artificial definition. For example even the simple three-termed sequence $\langle a, b, c \rangle$ is on this definition construed as the monstrosity

$$\{\{\{\{a\}, \{a, b\}\}\}, \{\{\{a\}, \{a, b\}\}, c\}\}$$

and the complexity mounts astronomically as our sequences become longer and longer. There are two other objections one might make to this 'analysis'. One is that it cannot be applied to infinite sequences (without infringing the axiom of foundation) though finite ordinals can perfectly well be applied to some infinite sequences, for example progressions. Another is that to adopt this analysis would be to commit oneself to some general axioms for set theory, which seems to be more than we need to do at this stage. For these axioms will evidently not be of the simple kind which can be construed innocuously along the lines of Russell's 'no-class' theory, because it is clear that we shall want

---

[22] A more economical version (not assuming that there is any one sequence which is *the* sequence of Φ's generated by $\mathfrak{R}$) would be: α is the nth term of any sequence such that (i) all its terms are Φ's, (ii) each term bears $\mathfrak{R}$ to all succeeding terms, and (iii) every Φ such that α does not bear $\mathfrak{R}$ to it is a term of the sequence.

the type of a sequence to depend only on the type of its terms whereas on this proposal the type of the sequence will depend also on its length, unless we say that if a and b are of the same type then $\{a, b\}$ is also of that same type. And if we do say the latter, then the set theory must make a real assumption about the existence of the set $\{a, b\}$[23]. Now if we do intend to make such realistic assumptions, then, since our immediate purpose is with sequences rather than sets in general, it would seem to be more to the point to try to formulate the required assumptions specifically as assumptions about sequences rather than to begin with assumptions about sets and then to invoke this highly artificial 'analysis' of sequences in terms of sets, which will anyway limit us to finite sequences.

The most general characteristics of sequences could perhaps be formulated in this way. First, there are *unit* sequences, i.e. sequences with just one term. Indeed for anything x there is a unit sequence with x as its only term, and there is only one such sequence. If we represent the unit sequence with x as its only term by $\langle x \rangle$, and use α, β, . . ., temporarily as variables for sequences, this is to say that the operation which produces $\langle x \rangle$ from x is always performable, always yields a sequence α, and is a one–one operation. Therefore we may lay down as our first two axioms

(1) $(\exists \alpha)(\alpha = \langle x \rangle)$

(2) $\langle x \rangle = \langle y \rangle \rightarrow x = y$.

Next, any two sequences may be combined to yield a third sequence which is obtained by writing the one after the other without a break, i.e. there is an operation, say *concatenation*, which may be performed on any two sequences β and γ to yield a third sequence which we shall write as $\beta^\frown \gamma$. This operation is not commutative but it is associative, so we may lay down as our next two axioms

(3) $(\exists \alpha)(\alpha = \beta^\frown \gamma)$

(4) $(\alpha^\frown \beta)^\frown \gamma = \alpha^\frown(\beta^\frown \gamma)$.

So far our axioms do not distinguish sequences from sets, but

[23] I mean, we cannot paraphrase remarks about $\{a, b\}$ *à la* Russell, as remarks about the predicate '... $= a \lor ... = b$'.

the next two do. The first is perhaps the simplest instance of the principle that sequences cannot be the same unless they are the same length, *viz.*

(5) $\langle x \rangle \neq \alpha^\frown \beta$

and the second reflects the fact that sequences cannot be the same unless their terms are in the same order, *viz.*

(6) $\alpha^\frown \beta = \gamma^\frown \delta \rightarrow (\alpha = \gamma \ \& \ \beta = \delta) \lor (\exists \theta)(\alpha = \gamma^\frown \theta \ \& \ \theta^\frown \beta = \delta)$

$\lor \ (\exists \theta)(\gamma = \alpha^\frown \theta \ \& \ \theta^\frown \delta = \beta)$.

We may consider axioms (1)–(6) as giving the most *basic* properties of sequences, though it would hardly be reasonable to say that they provide by themselves an adequate axiomatization (see further pp. 54 *et seq.*). They are, however, sufficient for our immediate purposes.

First we may note that we can on this basis introduce the usual notation for finite sequences, defining

$$\langle x, y \rangle \quad \text{for} \quad \langle x \rangle^\frown \langle y \rangle$$

and more generally

$$\langle \ldots, y \rangle \quad \text{for} \quad \langle \ldots \rangle^\frown \langle y \rangle.$$

We may then prove (using axioms (2), (5), and (6)) the characteristic property of finite sequences:

$$\langle x, y \rangle = \langle z, w \rangle \rightarrow x = z \ \& \ y = w$$

and more generally

$$\langle \ldots, y \rangle = \langle \text{---}, w \rangle \rightarrow \langle \ldots \rangle = \langle \text{---} \rangle \ \& \ y = w.$$

We may also introduce the properties of sequences that we relied on in our explanation of the nth term of a sequence. We may put

$$\alpha \ begins \ \beta \quad \text{for} \quad \alpha = \beta \lor (\exists \gamma)(\alpha^\frown \gamma = \beta)$$

$$\alpha \ ends \ in \ \beta \quad \text{for} \quad \alpha = \beta \lor (\exists \gamma)(\alpha = \gamma^\frown \beta).$$

Then 'α is an initial segment of β' was defined as

$$\alpha < \beta \quad \text{for} \quad \alpha \ begins \ \beta \ \& \ \sim\!\beta \ begins \ \alpha.$$

It is quite easy to show (using axioms (3), (4), and (6)) that

(i) $\beta < \gamma \ \& \ \gamma < \delta \rightarrow \beta < \delta$

(ii) $\sim(\beta < \gamma \ \& \ \gamma < \beta)$

(iii) β *begins* $\alpha \ \& \ \gamma$ *begins* $\alpha \ \& \ \delta$ *begins* α

$\rightarrow (\beta < \gamma \rightarrow (\delta < \gamma \lor \beta < \delta))$

which is to say that the relation $<$ weakly orders the sequences which begin α. Therefore, as suggested, we may define

$$x \text{ is the } n^{\text{th}} \text{ term of } \alpha$$

for[24]

$$(\exists\beta)(\beta \text{ ends in } \langle x \rangle \ \& \ n^{\text{th}}(\beta, \text{ begins } \alpha, <) \ \&$$

$$(\forall\gamma\delta)(\gamma \text{ begins } \alpha \ \& \ \delta \text{ begins } \alpha \ \& \ \beta \not< \gamma \ \& \ \beta \not< \delta$$

$$\rightarrow (\gamma = \delta \lor \gamma < \delta \lor \delta < \gamma))).$$

This therefore completes the account of how the use of ordinals applying to the term of a sequence may be introduced as a secondary use, taking as primary the use of ordinals in conjunction with ordering relations.

It is also easy to see how, if axioms (1)–(6) may be taken for granted, we may instead introduce ordinals applying to the terms of a sequence as the primary use of ordinals. The simplest procedure here is probably to proceed by way of the notion of the *length* of a finite sequence. We may evidently lay down the recursive equivalences

α *is of length* $1 \leftrightarrow (\exists x)(\alpha = \langle x \rangle)$

α *is of length* $n' \leftrightarrow (\exists\beta)(\beta$ *is of length* $n \ \& \ (\exists x)(\alpha = \beta^{\frown}\langle x \rangle))$.

If we regard these as adequately specifying the predicates of sequences '... *is of length* n', we may then introduce ordinals by putting

$$x \text{ is the } n^{\text{th}} \text{ term of } \alpha$$

for

$$(\exists\beta)(\beta \text{ begins } \alpha \ \& \ \beta \text{ is of length } n \ \& \ \beta \text{ ends in } \langle x \rangle).$$

We can if we like use the same method as before to replace

---

[24] Note that the last clause $(\forall\gamma\delta)(...)$ in this definition is condition C of p. 38. If we had introduced n^{th} by means of the strong recursive equivalences A, this clause would of course be superfluous.

the recursive equivalences by a definition. Using Φ, Ψ, ...,
now as variables for predicates of sequences we define

$$\Phi'\alpha \quad \text{for} \quad (\exists\beta)(\Phi\beta \;\&\; (\exists x)(\alpha = \beta^\frown\langle x\rangle))$$

$$\Phi S\Psi \quad \text{for} \quad (\forall\alpha)(\Phi\alpha \leftrightarrow \Psi'\alpha)$$

and then we regard our length-predicates as introduced by
contextual definition so that

$$(\forall n)(-is\ of\ length\ n-)$$

abbreviates

$$(\forall\phi)(\phi * S(is\ of\ length\ 1) \rightarrow (-\phi-)).$$

Using axioms (1)–(6) for sequences, we may then without difficulty establish Peano's postulates for the predicates '... *is of
length n*', and thereby for the ordinals '... *is the nth term of* ...'.
But this, it must be admitted, can hardly be regarded as a satisfactory *logicist* construction because of course the axioms for
sequences are presupposed, and these do not at once recommend
themselves as 'purely logical' truths, or indeed as 'analytic',
since they do contain some very noticeable existential assumptions. So the question arises: can our axioms for sequences
be either eliminated or at any rate justified by a suitable
analysis of the notion of a sequence?

So far as I can see, the right answer to this question is in the
negative, though some headway could be made if we were to
limit our attention to *finite* sequences. Suppose we begin by
assuming that we do already have the natural numbers available in some form or other, and therefore that we already have
a paradigm sequence, containing no repetitions, consisting of the
natural numbers in their natural order. Now where α is a finite
sequence it is clear that there must be a relation which holds
between each term of α and each of the natural numbers up to
some given number, namely the relation of occupying the same
positions in α and in the paradigm sequence. Since the paradigm
sequence of natural numbers contains no repetitions, this relation will never relate different terms of α to the same natural
numbers; further if the relation relates some natural number n
to a term of α, then it must also relate all natural numbers less
than n to a term of α; finally, since α is finite, not all natural

numbers will be related by this relation. Thus there will be a non-null relation, say **f**, from natural numbers to the terms of α, such that

(i) $(\forall n)(\forall xy)(\mathbf{f}(n, x) \ \& \ \mathbf{f}(n, y) \rightarrow x = y)$

(ii) $(\forall n)((\exists x)(\mathbf{f}(n', x)) \rightarrow (\exists x)(\mathbf{f}(n, x)))$

(iii) $(\exists n)(\forall x)(\sim\!\mathbf{f}(n, x))$.

The connection between this relation and the sequence α is of course that

(iv) $(\forall n)(\forall x)(x \ is \ the \ n^{\text{th}} \ term \ of \ \alpha \leftrightarrow \mathbf{f}(n, x))$

Conversely, wherever we have a non-null relation **f** satisfying these conditions, it is clear that α must be a finite sequence. In a reductive mood, therefore, we may *identify* a finite sequence with a non-null relation satisfying conditions (i)–(iii), for the sequence is fully identified as the sequence α such that (iv) holds.

Suppose, therefore, that in our axioms (1)–(6) for sequences we interpret the variable α as being simply a variable for the non-null relations satisfying (i)–(iii) above. At the same time we shall interpret $\langle x \rangle$ as standing for the relation which relates 1 to x and nothing else to anything else, and we shall interpret $\alpha^\frown\beta$ as standing for the relation which is the logical sum of the relation α and the relation which results from the relation β upon exchanging each number n related by β for the number $n + m$, where m is the highest of the numbers related by α. It is easily checked that under this interpretation our six axioms for sequences become deducible simply from the properties of the paradigm sequence of natural numbers. So by this method we have apparently succeeded in removing the axiomatic status of our six axioms. Unfortunately, however, this does not get us much further forward with our original design.

One difficulty is that I began this reinterpretation of the axioms by supposing that we already had the natural numbers available in some form as a paradigm sequence without repetitions. However, if our ultimate goal is to be able to introduce numbers for the first time as ordinals applying to sequences, it hardly helps to begin by using numbers in the analysis of sequences. But it seems that this difficulty could be overcome. It is fairly easily seen that any sequence which is a progression and contains no repetitions would be satisfactory as our paradigm sequence, and it is not necessary to suppose that it is in any sense a progression *of the numbers*. Nor in fact does it seem necessary to start by assuming the existence of any particular paradigm progression. We might instead begin with general conditions for being a progression without repetitions (for example, that Peano's postulates should be satisfied), treat our six axioms as conditional in form, prefaced by the hypothesis 'if there is a progression then . . .', and then go on to *prove* that the predicates '. . . *is of length n*' do constitute a progression without repetitions by applying them to sequences of predicates of this same form. However, I shall not pursue this suggestion in any detail, because there seems to be a more fundamental objection.

This is that we have only been able to interpret the axioms as having any sort of claim to be truths of logic by restricting them to *finite* sequences. But we have already pointed out that finite ordinals are also applicable to some infinite sequences, for example to progressions, and it was partly to accommodate this fact that we abandoned the first set-theoretical analysis of finite sequences in favour of some more general axioms. Now it seems that we are back where we started, because we have only offered to establish these axioms in the special case where all the sequences concerned are finite. It is of course true that the reductive analysis of a finite sequence as simply a non-null relation satisfying our conditions (i)–(iii) could easily be extended to cover progressions also, simply by dropping clause (iii) from the conditions. However, if we do drop clause (iii), thus allowing the variable α to include progressions within its range, then under the interpretation we have given axiom (3) no longer holds. If α is itself a progression, then there will be no relation β from natural numbers to other things such that $\beta = \alpha^\frown \gamma$. One response to this difficulty might be to rewrite axiom (3) with an explicit limitation to finite sequences[25], but (a) this limitation seems entirely *ad hoc*, for there is no reason to suppose that axiom (3) holds only for finite sequences, and

[25] 'Finite sequence' is defined on p. 55.

(b) the original difficulty is really not avoided thereby. For the original difficulty was that finite ordinals can be used of terms of sequences which are not themselves finite sequences, so if we confine our analysis to finite sequences we shall not be doing full justice to the notion of a finite ordinal. It will not meet this difficulty to extend our analysis to cover progressions as well, unless finite ordinals can sensibly be applied only to sequences which are either finite or progressions. And this is not so. For one thing, it is clear that finite ordinals may sensibly (and truly) be applied to the initial terms of sequences which are longer than progressions, e.g. to the sequence (of type $\omega + \omega$) consisting of all the even numbers in their natural order and then all the odd numbers in their natural order. For another thing, finite ordinals may sensibly (and falsely) be applied to sequences which do not even have a first term at all, for example the sequence (of type ω^*) consisting of all the negative numbers in their natural order. For it is true of this sequence that it has no first term, and indeed no n^{th} term for any n, and that is to say that the predicates '$(\exists x)(x$ *is the* n^{th} *term of* $\ldots)$' are significantly, and falsely, applicable to it. So the basic objection here is that finite ordinals are significantly applicable to *any* sequence, and there is no prospect of a reductive analysis of *all* sequences on the lines we have sketched. I conclude that for *logicist* purposes we cannot take the application of numbers to sequences as primary just because the notion of a sequence cannot itself be introduced without non-logical assumptions.

To summarize the argument of this section, we began by taking as primary the use of ordinals in conjunction with (weakly) ordering relations, the basic context being taken to be 'α is the nth Φ under $\mathfrak{R}$'. We found it possible to introduce ordinals in this use without presupposing an understanding of numbers in any other role, though of course it is also possible to define this use in terms of numerical quantifiers or in terms of numerical powers of relations. However, I pointed out that ordinals are also used of the terms of sequences which are not (in the ordinary sense) ordered by an ordering relation. From our original starting point we were able to go on to explain this as a secondary use of ordinals. For this purpose we must of course invoke some properties of sequences, just to show what the relevant ordering relation is, but in the present context that is unobjectionable. For on this first approach we do not need to assume or prove any properties of sequences in order to explain the notion of an ordinal in the first place, or to establish the properties of ordinals. The application of ordinals to sequences is treated as only one of a number of special applications, and the fact that we have to make some assumptions about sequences in order to explain this application properly is no more damaging than it would be damaging to our account of numerical quantifiers if we had to make some special assumptions about Homeric gods in order to explain the relevant identity-relation to be used in elucidating 'there are just twelve Homeric gods'.

On the other hand it seems quite plausible to say that the conception of *order* which allows us to think even of a repetitive sequence as an ordering is more fundamental, because more general, than the special case where the order can be seen as generated by an ordering relation. One sign of this is that when this more general conception of an order is available we can easily explain numerical powers of relations in terms of ordinals, since a bears the nth power of a relation R to b if and only if there is a sequence (possibly containing repetitions) such that a is its first term and b its nth term, and each term of the sequence bears R to the next. However, it is not so simple to define numerical powers of relations in terms of ordinals which are used only in conjunction with ordering relations[26]. We therefore raised the question of whether our direction of explanation could be reversed, taking the more general conception of order as fundamental and introducing our finite ordinals with this as their primary application. The only noticeable difficulty here is the difficulty of giving a suitable analysis of this conception of order. If we allow ourselves to make primitive use of concepts such as that of the *concatenation* of two sequences, with appropriate axioms to govern it, this may be possible. However, the concept in question would not normally be reckoned as itself a 'purely logical' concept, and I see no way of defining it in terms of accepted 'purely logical' concepts, nor therefore of producing any useful justification of the axioms governing it. So this approach to numbers seems to require

---

[26] A somewhat roundabout definition will be given shortly.

a non-logical starting point to provide the requisite conception of an order, but once started the approach runs
smoothly enough, and in view of the ease with which it allows
one to define other uses of the natural numbers there may be
something to be said for it.

Finally I should perhaps point out that the idea of 'analysing'
a sequence as a correlation between natural numbers and the
terms of the sequence, though it does not do justice to the
notion of a sequence, can nevertheless be properly applied to
define some uses of numbers in terms of others. For example we
have just noted that numerical powers of relations can easily be
defined in terms of ordinals applied to sequences, and we can
properly apply the suggested 'analysis' here since only *finite*
sequences need be invoked. That is, *however* the natural numbers
are first introduced, we may always explain powers of relations
by saying that a bears the nth power of the relation R to b if and
only if there is a correlation $\mathbf{f}$ between our numbers and other
things which satisfies conditions (i)–(iii) of p. 49 and is also
such that

(iv) $(\forall n)(\forall xy)(\mathbf{f}(n, x) \,\&\, \mathbf{f}(n', y) \rightarrow xRy)$

(v) $\mathbf{f}(1, a) \,\&\, \mathbf{f}(n', b)$

This is in fact the usual method of defining numerical powers of
relations when cardinal numbers are taken as the primary
numbers[27]. We could if we wished introduce ordinals in the
same way, saying that a is the nth so-and-so if and only if
$\mathbf{f}(n, a)$ for some suitable correlation $\mathbf{f}$, though in this case, as
we have seen, it is quite a problem to see what 'so-and-so' represents in this schema and therefore what counts as a 'suitable'
correlation. Again we could if we wished introduce finite cardinals by saying that there are just n things x such that Fx if
and only if there is a relation $\mathbf{f}$ between (some version of) the
natural numbers and predicates such that

(i) $\mathbf{f}(n, F)$

(ii) $(\forall n)((\exists G)(\mathbf{f}(n', G)) \rightarrow (\exists G)(\mathbf{f}(n, G)))$

(iii) $(\forall G)(\mathbf{f}(1, G) \rightarrow (\exists x)(\forall y)(Gy \leftrightarrow y = x))$

---

[27] See e.g. Quine (1951, § 47) for a minor variation.

(iv) $(\forall n)(\forall GH)(\mathbf{f}(n', G)\ \&\ \mathbf{f}(n, H) \rightarrow (\exists x)(Gx$

$\&\ (\forall y)(Hy \leftrightarrow Gy\ \&\ y \neq x)))$.

I do not suggest that this would be a very natural way of introducing cardinals, but clearly it will always be possible whatever form the first introduction of numbers has taken.

It is not a very surprising fact about the natural numbers that there are all sorts of different ways of explaining one use of natural numbers in terms of others. Nor would this fact be of any noticeable significance if the starting point was somehow fixed and we were confronted with a choice only over the ways of introducing secondary uses. However, it is of some significance that the starting point is not fixed, and that we may choose quite different uses of the natural numbers to take as the primary use. But I postpone further comment on this point to Chapter 4. In the next chapters I shall discuss yet another of the uses of natural numbers, and this time it is a use which they share with rational numbers and with real numbers. Before I come to this I end the present chapter with a section which is somewhat a of digression to my main theme, but is needed to introduce a principle we shall later be making use of.

4. Progressions

The six axioms for sequences which I suggested in the last section were

(1) $(\exists \alpha)(\alpha = \langle x \rangle)$

(2) $\langle x \rangle = \langle y \rangle \rightarrow x = y$

(3) $(\exists \alpha)(\alpha = \beta ^\frown \gamma)$

(4) $\alpha ^\frown (\beta ^\frown \gamma) = (\alpha ^\frown \beta) ^\frown \gamma$

(5) $\langle x \rangle \neq \alpha ^\frown \beta$

(6) $\alpha ^\frown \beta = \gamma ^\frown \delta \rightarrow ((\alpha = \gamma\ \&\ \beta = \delta) \vee (\exists \theta)(\alpha = \gamma ^\frown \theta$
$\&\ \theta ^\frown \beta = \delta) \vee (\exists \theta)(\gamma = \alpha ^\frown \theta\ \&\ \theta ^\frown \delta = \beta))$.

If we intended to restrict our attention to *finite* sequences these axioms would be sufficient provided we add a further axiom to effect the restriction. For we might observe that it is characteristic of finite sequences that they can all be generated by applying only axioms (1) and (3) above, and hence we may ensure that the variables range only over finite sequences by adding as our extra axiom a principle of induction

$$(\forall x)(\Phi\langle x\rangle) \ \& \ (\forall\alpha\beta)(\Phi\alpha \ \& \ \Phi\beta \rightarrow \Phi(\alpha^\cap\beta)) \rightarrow (\forall\alpha)(\Phi\alpha).$$

All desired results for finite sequences would then be forthcoming.

On the other hand, the satisfactoriness of this system does depend upon the fact that our extra axiom is a very strong one, and there are other ways of representing the restriction to finite sequences which would not suffice as the sole further axiom. For example we could define

α *is well-ordered*   for $(\forall\beta)(\alpha$ *ends in* $\beta\rightarrow(\exists x)(\langle x\rangle$ *begins* $\beta))$

α *is* $^\cup$*well-ordered*   for $(\forall\beta)(\beta$ *begins* $\alpha\rightarrow(\exists x)(\beta$ *ends in* $\langle x\rangle))$

α *is finite*   for α *is well-ordered* $\&$ α *is* $^\cup$*well-ordered.*

From the inductive axiom just suggested we can of course deduce that, in the sense just defined,

$$(\forall\alpha)(\alpha \ is \ finite).$$

However, the converse deduction is not possible. So if we had instead adopted this latter as our extra axiom the system would be seriously incomplete.

It may be of interest to illustrate this by considering how we might attempt to deduce the required induction principle, though for convenience I shall change to a slightly different version of that principle (easily seen to be equivalent to the original version), namely

$$(\forall x)(\langle x\rangle \ begins \ \alpha \rightarrow \Phi\langle x\rangle)$$
$$\& \ (\forall\beta)(\forall x)(\Phi\beta \ \& \ \beta^\cap\langle x\rangle \ begins \ \alpha \rightarrow \Phi(\beta^\cap\langle x\rangle))$$
$$\rightarrow \Phi\alpha.$$

Now the assumption that α is finite, so that every final segment of α has a first term, allows us to show that

$$(\forall\beta)(\beta \ begins \ \alpha \rightarrow \beta = \alpha \ \vee \ (\exists x)(\beta^\cap\langle x\rangle \ begins \ \alpha)).$$

Therefore if we suppose (for *reductio ad absurdum*) that the two antecedents to the inductive principle are true and the conclusion false we evidently deduce

$$(\exists x)(\langle x \rangle \ begins \ \alpha \ \& \ \Phi\langle x \rangle) \ \&$$

$$(\forall\beta)(\beta \ begins \ \alpha \ \& \ \Phi\beta \rightarrow (\exists x)(\beta^\cap\langle x \rangle \ begins \ \alpha \ \& \ \Phi(\beta^\cap\langle x \rangle))).$$

This is to say, from an intuitive point of view, that α has infinitely many finite initial segments with the property Φ, each one term longer than the last. It would seem, then, that we should be able to complete our proof by showing that there must be an initial segment of α that has no last term, namely the one that is the 'least upper bound' of all the finite initial segments with the property Φ. If so, then α is not finite according to our definition, and the proof is completed. Of course the catch is that our present axioms do not allow us to prove the existence of this desired 'least upper bound'.

We might therefore consider adding an appropriate existential postulate to our present axioms. It turns out, however, that on the obvious way of defining 'least upper bound' we are still not able to prove that the least upper bound in question has no last term, and perhaps the simplest resolution of the difficulty is to begin by adding a further axiom of a general nature

(7) $\alpha \neq \langle x \rangle \ \& \ (\forall\gamma)(\forall y)(\gamma^\cap\langle y \rangle \ begins \ \alpha \rightarrow \gamma^\cap\langle y \rangle \ begins \ \beta)$

  $\rightarrow \alpha \ begins \ \beta$

and then framing the existential postulate as[28]

(8) $(\exists\alpha)(\Phi\alpha) \ \& \ (\exists\beta)(\forall\alpha)(\Phi\alpha \rightarrow \alpha \ begins \ \beta) \rightarrow$

  $(\exists\beta)(\forall\gamma)(\forall y)(\gamma^\cap\langle y \rangle \ begins \ \beta \leftrightarrow (\exists\alpha)(\gamma^\cap\langle y \rangle \ begins \ \alpha \ \& \ \Phi\alpha)).$

Axioms (7) and (8) no doubt hold for infinite sequences as well as for finite ones, and with their aid we can certainly complete the proof that induction holds for finite sequences. In fact one can go quite a long way on the basis of axioms (1)–(8) without any restriction to finite sequences—for example one can prove appropriate principles of transfinite induction for sequences that are well ordered in the sense just defined. But that is perhaps rather by the way, for it is clear that axioms (1)–(8) are still far too weak as a general theory of sequences, because in the general

[28] With axioms (7) and (8) compare axioms (P1) (from right to left) and (P2) on p. 119.

theory one would wish to provide for the existence of infinite sequences as well as finite ones, but axioms (1)–(8) do not yet allow us to prove that there are any infinite sequences. It is clearly a difficult and important problem to determine what infinite sequences we should want to have as existing, and to provide for their existence without, for example, falling foul of the Burali–Forti paradox. I do not intend to pursue this problem, nor therefore to try to improve our general axioms for sequences any further, but at least it seems clear that we shall want to provide for the existence of *progressions*, and that is the topic of this section.

Progressions can be defined in several ways within our present terminology; for example we might put

$$\alpha \ is \ a \ progression$$

for[29]

$$\sim(\alpha \ is \ finite) \ \& \ (\forall\beta\gamma)(\alpha = \beta^\frown\gamma \to \beta \ is \ finite).$$

However, our question was rather how to provide for their existence. It is pretty clear to intuition that a progression is determined if each of its finite initial segments is determined, when of course the condition to be met is that each such initial segment is a continuation of the previous one. It would seem reasonable to say, then, that wherever we have a denumerable set of finite sequences, each being a continuation of the previous one, those sequences can be amalgamated to form a progression. Therefore an appropriately general principle providing for the existence of progressions might be

$$(\exists x)(\Phi\langle x\rangle) \ \& \ (\forall\alpha)(\Phi\alpha \to (\exists x)(\Phi(\alpha^\frown\langle x\rangle))) \to$$
$$(\exists\beta)(\beta \ is \ a \ progression \ \& \ (\forall\alpha\gamma)(\beta = \alpha^\frown\gamma \to \Phi\alpha)).$$

This is in fact a rather generous principle. It would be consistent with the previous remarks to say that we are entitled to assume the existence of a progression only when we have a *unique* set of finite sequences specified as its initial segments, so that we should rather say (using strong numerical quantifiers)

$$(1x)(\Phi\langle x\rangle) \ \& \ (\forall\alpha)(\Phi\alpha \to (1x)(\Phi(\alpha^\frown\langle x\rangle))) \to$$
$$(\exists\beta)(\beta \ is \ a \ progression \ \& \ (\forall\alpha\gamma)(\beta = \alpha^\frown\gamma \to \Phi\alpha)).$$

[29] Note that according to this definition progressions may contain repetitions.

In the theory of sequences this weaker principle would suffice for many purposes—for example it could replace axiom (8) in the proof that induction holds for finite sequences—and if we adopt the first and stronger principle instead then we are in effect subscribing to a form of the axiom of choice. My purpose for the remainder of this section is to discuss this form of the axiom of choice, because I shall be making some use of it in the next chapter. It is better known in a different form under the title 'Principle of Dependent Choices', and though I shall continue to call it a principle providing for the existence of progressions it will be useful first to disentangle it from the special notation (and axioms) we have been employing for the unanalysed notion of a sequence.

The effect of adopting the more generous principle for the existence of progressions is to commit oneself to the view that the antecedent of the more generous principle entails the antecedent of the less generous one in the sense that

$$(\exists x)(\Phi\langle x\rangle) \mathbin{\&} (\forall\alpha)(\Phi\alpha\to(\exists x)(\Phi(\alpha^\cap\langle x\rangle))) \to$$
$$(\exists\Psi)((\forall\alpha)(\Psi\alpha\to\Phi\alpha) \mathbin{\&} (1x)(\Psi\langle x\rangle) \mathbin{\&} (\forall\alpha)(\Psi\alpha\to(1x)(\Psi(\alpha^\cap\langle x\rangle)))).$$

(To see this, simply take Ψ as applying to just the finite initial segments of the progression which the more generous principle requires to exist.) In this revised version we have only finite sequences to deal with, so we may without loss employ our old technique of replacing finite sequences by relations between natural numbers and other things (p. 49). However, a straightforward application of this technique is somewhat unwieldy in the present instance, and we can streamline the procedure a little. Where it is true that for some predicate Φ of sequences

(i) $(\exists x)(\Phi\langle x\rangle) \mathbin{\&} (\forall\alpha)(\Phi\alpha\to(\exists x)(\Phi(\alpha^\cap\langle x\rangle)))$

there it will also be true that there is a relation $\mathbf{f}$ from numbers to terms of sequences and a relation R between terms of sequences such that

(ii) $(\exists x)(\mathbf{f}(1, x)) \mathbin{\&} (\forall n)(\forall x)(\mathbf{f}(n, x)\to(\exists y)(xRy \mathbin{\&} \mathbf{f}(n', y)))$

for $\mathbf{f}$ may be taken as the relation which holds between the number n and the last term of any n-termed sequence α such that $\Phi\alpha$, and R may be taken as the relation which holds between x and y if and only if there is some finite sequence α which ends in $\langle x \rangle$ and is such that $\Phi\alpha$ & $\Phi(\alpha^\frown\langle y \rangle)$. Conversely wherever there are relations $\mathbf{f}$ and R satisfying (ii) then there is also a predicate Φ of sequences satisfying (i), for we have only to understand Φ as true of the finite sequences α such that $\mathbf{f}(n, x)$ holds wherever x is the nth term of α and each term of α bears R to the next. If in the same way we rephrase the consequent

(iii) $(\exists\Psi')((\forall\alpha)(\Psi'\alpha \rightarrow \Phi\alpha)$ & $(1x)(\Psi'\langle x \rangle)$
    & $(\forall\alpha)(\Psi'\alpha \rightarrow (1x)(\Psi'(\alpha^\frown\langle x \rangle)))).$

in a form appropriate to our new version (ii) of the antecedent, it is easy to see by the same reasoning that it becomes

(iv) $(\exists\mathbf{g})((\forall n)(\forall x)(\mathbf{g}(n, x) \rightarrow \mathbf{f}(n, x))$ & $(1x)(\mathbf{g}(1, x))$
    & $(\forall n)(\forall x)(\mathbf{g}(n, x) \rightarrow (1y)(xRy$ & $\mathbf{g}(n', y)))).$

On the assumption, roughly put, that all finite sequences exist, one can thus show that the principle that (i) implies (iii) is interchangeable with the principle that (ii) implies (iv), and from the point of view of logical analysis the second version has the advantage that it does not explicitly invoke the (non-logical) concept of a sequence.

At this point it will be convenient to make some notational alterations. In what I have said so far I have been regarding the variables x and y, as they occur in this principle, as ranging over the sort of things that can be terms of sequences, but presumably items of any type can be terms of sequences, and anyway I do not intend there to be any type-restrictions on these variables. It may be better, then, to return to the style of variable that has already been introduced as type-neutral, so x will now be replaced by α (which is no longer a variable for sequences) and R will be replaced by $\mathfrak{R}^{30}$. I shall also change the use of $\mathbf{f}$ and $\mathbf{g}$. It is clear that any relation $\mathbf{g}$ from natural numbers to other things which satisfies the condition

$(1\alpha)(\mathbf{g}(1, \alpha))$ & $(\forall n)(\forall\alpha)(\mathbf{g}(n, \alpha) \rightarrow (1\beta)(\alpha\mathfrak{R}\beta$ & $\mathbf{g}(n', \beta)))$

---

[30] I always use $\mathfrak{R}$ as restricted to *homogeneous* relations.

must be, or at least contain, a many–one relation $\mathbf{f}$ with every natural number in its domain, i.e. a relation $\mathbf{f}$ such that

$$(\forall n)(\exists \alpha)(\mathbf{f}(n, \alpha))$$

$$(\forall n)(\forall \alpha \beta)(\mathbf{f}(n, \alpha) \ \& \ \mathbf{f}(n, \beta) \rightarrow \alpha = \beta).$$

This relation $\mathbf{f}$ can therefore be regarded as a *function* from the natural numbers to other things, and we may with advantage use the usual notation for functions by which $\mathbf{f}(n)$ stands for the sole α such that $\mathbf{f}(n, \alpha)$. So for simplicity I shall henceforth use $\mathbf{f}$ and $\mathbf{\dot{g}}$ as variables restricted to functions from the natural numbers to other things[31]. The occurrence of $\mathbf{f}$ in the antecedent to the principle thus becomes inappropriate, and I replace it by $\mathscr{A}$, with the intention that $\mathscr{A}(n, \alpha)$ shall stand in for any formula with free occurrences of n and α. In this revised notation the principle that (ii) above implies (iv) may be reformulated as the principle that

if $\quad (\exists \alpha)(\mathscr{A}(1, \alpha)) \ \& \ (\forall n)(\forall \alpha)(\mathscr{A}(n, \alpha) \rightarrow (\exists \beta)(\alpha \Re \beta \ \& \ \mathscr{A}(n', \beta)))$

then $\quad (\exists \mathbf{f})(\forall n)(\mathscr{A}(n, \mathbf{f}(n)) \ \& \ \mathbf{f}(n) \Re \mathbf{f}(n')).$

This last version of the principle is the one that I find most convenient to use hereafter, but it may be noted that the enunciation can certainly be simplified; we can strengthen the antecedent and weaken the consequent without in fact suffering any loss. First we may observe that it would make no difference if we added to the antecedent the further condition

$$(\forall nm)(\forall \alpha)(\mathscr{A}(n, \alpha) \ \& \ \mathscr{A}(m, \alpha) \rightarrow n = m).$$

In outline the reasoning is that if our original antecedent is satisfied then we may always satisfy the new stronger antecedent by considering not the items α such that $(\exists n)(\mathscr{A}(n, \alpha))$ but rather the finite sequences of such items in which each term bears $\Re$ to the next. Therefore we replace $\mathscr{A}$ by the relation which holds from the natural number n to the n-termed sequences of this sort, and this evidently is a one–many relation as required. We also replace $\Re$ by the relation which holds

---

[31] I here assume that this notion is sufficiently familiar to need no further explanation at this stage. For a more detailed discussion see the note appended to this chapter (p. 73).

between two such sequences when the second is a continuation of the first. The principle may then be applied, with the stronger antecedent now satisfied, to generate a progression of these sequences, and from the non-repetitive progression of sequences we may then return to the (possibly repetitive) progression of items in those sequences which was originally desired. We may note that since the reasoning involves only finite sequences we may use our earlier reductive technique (p. 49) to replace these by relations from numbers to other things and no special axioms about sequences will need to be assumed (though we do of course assume the type-neutrality of our principle). Thus we may always add the further antecedent

$$(\forall nm)(\forall \alpha)(\mathscr{A}(n, \alpha) \ \& \ \mathscr{A}(m, \alpha) \to n = m)$$

without affecting our principle in the slightest.

Second we may notice that the consequent may be weakened by altogether omitting the clause

$$\mathscr{A}(n, \mathbf{f}(n))$$

for we can recover this clause as follows. We begin by assuming the antecedent

(i) $(\exists \alpha)(\mathscr{A}(1, \alpha)) \ \& \ (\forall n)(\forall \alpha)(\mathscr{A}(n, \alpha) \to (\exists \beta)(\alpha \Re \beta \ \& \ \mathscr{A}(n', \beta)))$

so for some A we have

(ii) $\mathscr{A}(1, A)$.

It is clear that (where $*\Re$ is the improper ancestral of $\Re$) (i) and (ii) between them imply

(iii) $(\exists \alpha)(\mathscr{A}(1, \alpha) \ \& \ A*\Re\alpha) \ \& \ (\forall n)(\forall \alpha)(\mathscr{A}(n, \alpha) \ \& \ A*\Re\alpha$
$\to (\exists \beta)(A*\Re\alpha \ \& \ \alpha\Re\beta \ \& \ A*\Re\beta \ \& \ \mathscr{A}(n', \beta)))$.

We now take this thesis (iii) as a case of the antecedent to our principle, with $\mathscr{A}(n, \alpha) \ \& \ A*\Re\alpha$ in place of the original $\mathscr{A}(n, \alpha)$ and $A*\Re\alpha \ \& \ \alpha\Re\beta$ in place of the original $\alpha\Re\beta$. Thus we deduce that for some function $\mathbf{f}$

$$(\forall n)(A*\Re\mathbf{f}(n) \ \& \ \mathbf{f}(n)\Re\mathbf{f}(n'))$$

and hence in particular

(iv) $A*\Re\mathbf{f}(1) \ \& \ (\forall n)(\mathbf{f}(n)\Re\mathbf{f}(n'))$.

But it is quite straightforward to show that if A$\ast\Re$B then, for some finite number n, A$\Re^n$B, and hence that there is a function $\mathbf{g}$ such that

(v)　$\mathbf{g}(1) = \mathrm{A}\ \&\ \mathbf{g}(n') = \mathrm{B}\ \&\ (\forall m)(m \leqslant n \to \mathbf{g}(m)\Re\mathbf{g}(m'))$.

Bringing (iv) and (v) together we may therefore define a new function $\mathbf{h}$ such that, for all m,

$$\mathbf{h}(m) = \mathbf{g}(m) \qquad \text{if}\quad m \leqslant n$$

$$\mathbf{h}(m) = \mathbf{f}(m-n) \quad \text{if}\quad m > n.$$

This new function $\mathbf{h}$ is evidently so defined that

$$\mathbf{h}(1) = \mathrm{A}\ \&\ (\forall m)(\mathbf{h}(m)\Re\mathbf{h}(m')).$$

Therefore, using and discharging assumption (ii), we can infer as a first step that (i) implies

(vi)　$(\exists\mathbf{h})(\mathscr{A}(1, \mathbf{h}(1))\ \&\ (\forall n)(\mathbf{h}(n)\Re\mathbf{h}(n')))$.

The simplest way to complete the argument is now to make use of our earlier observation that the condition

$$(\forall nm)(\forall\alpha)(\mathscr{A}(n, \alpha)\ \&\ \mathscr{A}(m, \alpha) \to n = m)$$

may always be imposed without loss. For by using the clause $(\exists m)(\mathscr{A}(m, \alpha)\ \&\ \mathscr{A}(m', \beta)\ \&\ \alpha\Re\beta)$ in place of the original $\alpha\Re\beta$, and invoking the result just established that (i) implies (vi), we evidently infer

$(\exists\mathbf{h})(\mathscr{A}(1, \mathbf{h}(1))\ \&\ (\forall n)(\exists m)(\mathscr{A}(m, \mathbf{h}(n))\ \&$

$$\mathscr{A}(m', \mathbf{h}(n'))\ \&\ \mathbf{h}(n)\Re\mathbf{h}(n'))).$$

However, with the extra condition imposed on $\mathscr{A}$ this at once yields the desired result

$$(\exists\mathbf{h})(\forall n)(\mathscr{A}(n, \mathbf{h}(n))\ \&\ \mathbf{h}(n)\Re\mathbf{h}(n')).$$

This completes the argument to show that the first version of our principle stated above is in fact equivalent to this second and apparently weaker version:

if　　$(\exists\alpha)(\mathscr{A}(1, \alpha))\ \&\ (\forall n)(\forall\alpha)(\mathscr{A}(n, \alpha) \to (\exists\beta)(\alpha\Re\beta\ \&\ \mathscr{A}(n', \beta)))$

then　$(\exists\mathbf{f})(\forall n)(\mathbf{f}(n)\Re\mathbf{f}(n'))$.

With $\mathscr{A}$ now omitted from the consequent to the principle it is fairly easy to see that we can obtain a third version of the principle in which $\mathscr{A}$ is also omitted from the antecedent, and everything depends on the relation $\mathfrak{R}$. This is how the 'Principle of Dependent Choices' is usually stated, a standard version being[32]

if $(\exists\alpha\beta)(\alpha\mathfrak{R}\beta)\ \&\ (\forall\beta)((\exists\alpha)(\alpha\mathfrak{R}\beta)\rightarrow(\exists\gamma)(\beta\mathfrak{R}\gamma))$

then $(\exists\mathbf{f})(\forall n)(\mathbf{f}(n)\mathfrak{R}\mathbf{f}(n'))$.

For if the antecedent to this third version of the principle is satisfied we can always obtain the antecedent to the second by taking $\mathscr{A}(n,\alpha)$ as $A\mathfrak{R}^n\alpha$ for some A, and conversely if the antecedent to the second version of the principle is satisfied we can always obtain the antecedent to the third in the form

$(\exists\alpha\beta)(\alpha\mathfrak{R}\beta\ \&\ (\exists n)(\mathscr{A}(n,\beta))$

$\&\ (\forall\beta)((\exists\alpha)(\alpha\mathfrak{R}\beta\ \&\ (\exists n)(\mathscr{A})n,\beta)))\rightarrow(\exists\gamma)(\beta\mathfrak{R}\gamma\ \&\ (\exists n)(\mathscr{A}(n,\gamma))))$.

So these two versions are easily shown to be equivalent.

A fourth version of the principle is yet more simple to state, *viz.*[33]

if $(\forall\alpha)(\exists\beta)(\alpha\mathfrak{R}\beta)$

then $(\forall\alpha)(\exists\mathbf{f})(\mathbf{f}(1)=\alpha\ \&\ (\forall n)(\mathbf{f}(n)\mathfrak{R}\mathbf{f}(n')))$.

If we suppose the antecedent to the third version is satisfied, then for some A we shall have

$$(\forall\alpha)(\exists\beta)(A*\mathfrak{R}\alpha\rightarrow\alpha\mathfrak{R}\beta).$$

Hence by the fourth version of the principle

$$(\exists\mathbf{f})(\mathbf{f}(1=A\ \&\ (\forall n)(A*\mathfrak{R}\mathbf{f}(n)\rightarrow\mathbf{f}(n)\mathfrak{R}\mathbf{f}(n')))$$

from which the conclusion to the third version follows at once. Thus the fourth version implies the third. Conversely if the antecedent to the fourth version of the principle is satisfied then for any A it holds that

$$(\exists\alpha\beta)(A*\mathfrak{R}\alpha\ \&\ \alpha\mathfrak{R}\beta)$$

$$(\forall\beta)((\exists\alpha)(A*\mathfrak{R}\alpha\ \&\ \alpha\mathfrak{R}\beta)\rightarrow(\exists\gamma)(A*\mathfrak{R}\beta\ \&\ \beta\mathfrak{R}\gamma)).$$

[32] See Tarski 1948.
[33] I owe this version to Professor D. S. Scott.

Hence by the third version of the principle

$$(\exists f)(\forall n)(A*\Re f(n) \ \& \ f(n)\Re f(n'))$$

and so in particular

$$(\exists f)(A*\Re f(1) \ \& \ (\forall n)(f(n)\Re f(n'))).$$

From here we can use the method employed earlier (p. 62) to define a new function **h** satisfying

$$h(1) = A \ \& \ (\forall n)(h(n)\Re h(n'))$$

which is the consequent to the fourth version of the principle. Thus once again the third and fourth versions are interdeducible. There are of course yet other ways of stating the principle, but these will suffice. The first and fullest version is the one that I have found most convenient to use and it is also the one that relates most closely to our previous discussion of sequences and progressions, but the later versions certainly gain in simplicity of statement.

Now what are we to say of this principle providing for the existence of progressions? First, it follows of course from the full and unrestricted axiom of choice. To see this we have only to recall that the axiom of choice is equivalent to the thesis that every class can be well-ordered. For if we suppose that $(\forall \alpha)(\exists \beta)(\alpha \Re \beta)$, and if we also suppose that for any α we can well-order the things β such that $\alpha*\Re\beta$, then we can evidently define the required function **f** recursively by specifying that $f(1)$ is to be α and $f(n')$ is to be the earliest entity β in the well-ordering which is such that $f(n)$ bears $\Re$ to β[34]. However, it seems to me that the thesis that every class can be well-ordered should be treated with some suspicion. For example one is quite at a loss to envisage a well-ordering of even such a well behaved class as the class of all real numbers, not to mention other more unwieldy collections. So this justification for the principle is one that I would rather avoid.

An important difference between our principle and the full axiom of choice is that, intuitively speaking, our principle permits us only to make denumerably many selections, one for

---

[34] It is well known that such a recursive definition can always be replaced, if desired, by an explicit one: roughly, we explicitly define $f(n)$ to be the entity α such that $g(n) = \alpha$ for every function g satisfying the recursive equations.

each natural number, whereas the full axiom of choice has no such restriction. This is only half the story, however, for there is another well-known principle, weaker than ours, which permits denumerably many selections and which we can put in a form suitable for comparison with our first version as the principle that

if $(\forall n)(\exists \alpha)(\mathscr{A}(n, \alpha))$

then $(\exists \mathbf{f})(\forall n)(\mathscr{A}(n, \mathbf{f}(n)))$.

This principle is generally known as the 'Countable Axiom of Choice'. It is evidently deducible from our principle (e.g. by taking $\Re$ trivially as the universal relation), but the converse deduction is impossible[35]. For whereas this weaker principle simply says that if we have any way of associating things with the natural numbers we then may select one representative for each number, our stronger principle adds that we may make our selection in such a way that *each choice depends upon the previous one*, since the representative chosen for n' must be one which bears the appropriate relation (namely $\smile\Re$) to the representative chosen for n (hence of course the name 'Principle of Dependent Choices').

To speak in the language of what 'we can choose' is perhaps rather misleading here, for the principle should not be seen as concerned in any such straightforward way with human abilities. Here, then, is a rather different way of visualizing the principle, and we may again consider the first version and so assume

$$(\exists \alpha)(\mathscr{A}(1, \alpha)) \ \& \ (\forall n)(\forall \alpha)(\mathscr{A}(n, \alpha) \rightarrow (\exists \beta)(\alpha \Re \beta \ \& \ \mathscr{A}(n', \beta))).$$

Now imagine the things α such that $(\exists n)(\mathscr{A}(n, \alpha))$ arrayed in levels, with an item α on the nth level just in case $\mathscr{A}(n, \alpha)$[36] and with two items connected by an arrow wherever the one bears $\Re$ to the other. Our premise then assures us that there is at least one item on each level, and that every item on one level is connected by an arrow to some item on the next. The conclusion we wish to draw is that there is a 'path' in the diagram,

[35] For a proof see Jech 1973, pp. 122–31.
[36] For simplicity we may, as we have seen, assume further that

$$(\forall nm)(\forall \alpha)(\mathscr{A}(n, \alpha) \ \& \ (\mathscr{A}(m, \alpha) \rightarrow n = m),$$

so that each item appears only once in the array.

i.e. a series of arrows such that one begins where the other ends, which passes through every level; for the items lying on this path will be the items which are the values of the desired function. We may add that it would be quite legitimate to delete from the diagram those arrows (if any) which do not run from one level to the next, and then it is clear that *every* way of tracing a course along the arrows must mark out a path of the sort required. Therefore it seems entirely evident that there must be such a path, whether or not we can 'choose' one, but the fact remains that this is something we just have to assume: there is no way of deducing it from premises with a clear title to be 'purely logical'.

The fundamental reason why we cannot *deduce* the existence of a function **f** satisfying the condition $(\forall n)(\mathbf{f}(n)\Re\mathbf{f}(n'))$ is first that the function has infinitely many arguments and values, and second that we cannot provide a way of specifying the function that will work in every case. If the function had only finitely many arguments and values there would be no difficulty, for it is quite easy to prove (by induction) that for each finite number m, however large, there exists a function **f** such that $\mathbf{f}(n)\Re\mathbf{f}(n')$ holds for every number n up to m. This we can prove without being able to specify the function in any way, for example without knowing, for any n, which items α are such that $\mathscr{A}(n, \alpha)$. (Very roughly, one proves in effect that there must be a way of listing the values of the function, even though we do not know what that way is; but the existence of such an unspecified list can be proved only where the list is finite.) Alternatively, if the function could be specified we could of course prove that it existed, for example if (to revert to our illustration) we could say that one should always take the leftmost arrow and could guarantee that there would always be a leftmost arrow. But it is not possible to show that there always will be a way of specifying the function in every case.

We may compare here Russell's delightful illustration[37] of the millionaire

'who bought a pair of socks whenever he bought a pair of boots, and never at any other time, and who had such a passion for

See Russell 1919, pp. 126–7 (I have made some trivial alterations to Russell's text). As we shall see (p. 70), Russell's problem is not *quite* the same as ours.

buying both that at last he had denumerably many pairs of boots and denumerably many pairs of socks.'

Now

'The problem is: how many boots had he, and how many socks? One would naturally suppose that he had twice as many boots and twice as many socks as he had pairs of each, and therefore that he had denumerably many of each, since that number is not increased by doubling. . . . [But] to prove that a class has denumerably many terms it is necessary and sufficient to find some way of arranging its terms in a progression. There is no difficulty in doing this with the boots. The *pairs* are given as denumerably many, and therefore as the field of a progression. Within each pair, take the *left* boot first and the *right* second, keeping the order of the pair unchanged; in this way we obtain a progression of all the boots. But with the socks no such principle of selection suggests itself, and we shall have to choose *arbitrarily*, with each pair, which to put first; and an infinite number of arbitrary choices is an impossibility. Unless we can find a *rule* for selecting . . . we do not know that a selection is even theoretically possible.'

However, as Russell at once goes on to remark,

'Of course, in the case of objects in space, like socks, we always can find some principle of selection. For example, take the centres of mass of the socks: there will be points p in space such that, with any pair, the centres of mass of the two socks are not both at exactly the same distance from p; thus we can choose, from each pair, that sock which has its centre of mass nearer to p. But there is no theoretical reason why a method of selection such as this should always be possible.'

Now in fact the concrete applications of our principle that we shall have primarily in mind will always be to spatial (or temporal) objects, and methods of specifying the required functions will therefore be available. For example we shall invoke the principle to show that an infinite line contains a progression of mutually discrete parts, each (say) one inch long[38], and here the progression can evidently be constructed by starting with one such part and a direction in which the line is infinite, and then choosing always as our next part the one *adjacent* to

[38] Compare pp. 132–4.

the last in the given direction. When we are concerned with an infinite area containing a progression of mutually discrete areas, each (say) of one square inch, then if we make no assumption about the shape of the area the specification will have to be a little more complicated, but it can still be supplied. (For example, select any point within the area and any direction. We consider the finite sub-areas of the given area which are of one square inch and which are bounded by the edges of the area (if there are any) and circular arcs drawn with the selected point as centre. Our first sub-area is the circle or part of a circle of one square inch with the selected point as centre; our next is the

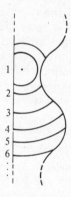

adjacent ring or partial ring of one square inch, and so on, unless a situation arises in which any subsequent ring would have two or more separated parts falling within the given area. In that case we choose the direction which is nearest to that initially selected and in which the part of the given area outside the last ring is infinite, and henceforth confine our construction to that part of the given area, ignoring the rest.) One would expect different specifications to be needed for different concrete cases, and I shall not attempt to spell out any general conditions under which specifications will be available, so from the formal point of view our principle for constructing progressions must be reckoned a new axiom[39].

I conclude this section with a brief discussion of how our

---

[39] The axiom is used three times on pp. 134–8.

version of the axiom of choice bears upon the cardinal 'denumerably many'. Now we say that there are denumerably many so-and-so's when the so-and-so's are *as many as* the natural numbers, and it is standard procedure to explain the idea of being 'as many as' in terms of the existence of one–one correlations. Of course this account of 'as many as' can be proved to be correct for finite cases, but it seems to me that it becomes progressively more dubious as we ascend to higher and higher orders of infinity (and this goes some way towards explaining the present state of uncertainty regarding the continuum hypothesis). However, that is rather by the way, for I do not here intend to consider any order of infinity other denumerability, and in this case the standard account is that there are denumerably many so-and-so's if and only if there exists a one–one correlation between the so-and-so's and the natural numbers, i.e. if and only if there is a one–one function from the natural numbers to the so-and-so's such that every so-and-so is a value of the function for some natural number as argument. Accordingly we define (using ∞ for 'denumerably many'[40])

$$(\infty\alpha)(\Phi\alpha)$$

for

$$(\exists\mathbf{f})((\forall nm)(\mathbf{f}(n) = \mathbf{f}(m) \to n = m) \ \& \ (\forall n)(\Phi(\mathbf{f}(n)))$$
$$\& \ (\forall\alpha)(\Phi\alpha \to (\exists n)(\alpha = \mathbf{f}(n)))).$$

It is quite easy to see that this definition can hardly be accepted unless at the same time we do accept a form of the axiom of choice which provides for the existence of the required functions. For example, in Russell's illustration of the boots and socks it is quite clear to intuition that there are as many boots as socks, and Russell can show quite satisfactorily from the standard definition that there are denumerably many boots, but cannot show without the axiom that there are denumerably many socks. I think we can only conclude that if the appropriate axiom is rejected then the intuitive idea of being as many as is not definable in this way, nor, presumably, in any other (extensional) way.

Now the form of the axiom of choice which is needed to

----

[40] I use ∞ rather than $\aleph_0$ only because I shall have nothing to say about other infinite cardinals $\aleph_\alpha$.

legitimize this definition is just the weak form known as the 'Countable Axiom of Choice', *viz.*

$$(\forall n)(\exists \alpha)(\mathscr{A}(n, \alpha)) \rightarrow (\exists \mathbf{f})(\forall n)(\mathscr{A}(n, \mathbf{f}(n))).$$

This is evidently sufficient to resolve the problem of the boots and socks, and it suffices for the proof of all the usual arithmetical properties of the cardinal 'denumerably many' when that is defined in the standard fashion. It is also sufficient, as Russell (1919, pp. 127–30) observed, to show that every infinite class contains a denumerable subclass, and I should like to dwell for a moment on this proposition. Continuing our earlier method for defining weak numerical quantifiers in terms of strong ones (p. 11) we can put

$$(\exists \infty \alpha)(\Phi \alpha) \quad \text{for} \quad (\exists \psi)((\infty \alpha)(\psi \alpha) \,\&\, (\forall \alpha)(\psi \alpha \rightarrow \Phi \alpha))$$

and we can given an independent account of 'infinitely many' as simply having no finite number, i.e.

$$(\forall n)(\sim(n\alpha)(\Phi \alpha))$$

or equivalently

$$(\forall n)(\exists n\alpha)(\Phi \alpha).$$

The thesis in question is then the thesis that

$$(\forall n)(\exists n\alpha)(\Phi \alpha) \rightarrow (\exists \infty \alpha)(\Phi \alpha)$$

and this, as observed, is provable from the countable axiom of choice above.

Now let us consider the possibility of extending this thesis to cover cases where the counting relation is other than non-identity. We may as before define

$$(\exists \infty_{\Re} \alpha)(\Phi \alpha)$$

for

$$(\exists \psi)((\exists \infty \alpha)(\psi \alpha) \,\&\, (\forall \alpha \beta)(\psi \alpha \,\&\, \psi \beta \,\&\, \alpha \neq \beta \rightarrow \alpha \Re \beta) \,\&\, (\forall \alpha)(\psi \alpha \rightarrow \Phi \alpha))$$

and then the thesis we have to consider is

$$(\forall n)(\exists n_{\Re} \alpha)(\Phi \alpha) \rightarrow (\exists \infty_{\Re} \alpha)(\Phi \alpha).$$

Now we find that where $\Re$ is taken as the paradigm sort of counting relation, namely the negation of an equivalence relation, we can already prove this extended thesis from the countable axiom of choice

$$(\forall n)(\exists \alpha)(\mathscr{A}(n, \alpha)) \rightarrow (\exists \mathbf{f})(\forall n)(\mathscr{A}(n, \mathbf{f}(n))).$$

In fact the converse holds too; that is, to adopt the countable axiom of choice is equivalent to adopting the above thesis for all relations $\Re$ which are the negations of equivalence relations. However, we cannot prove the thesis merely on the assumption that $\Re$ is irreflexive and symmetrical, for indeed it is not true for all such relations. A simple counter-example is obtained by taking

$\Phi\alpha$ as α is a line which is a part of the line l

$\alpha\Re\beta$ as α and β are equal and non-overlapping lines.

The antecedent to the thesis then states (roughly) that l can be divided into any finite number of equal and non-overlapping parts, which is true of any finite line l no matter how short, while the consequent states that l can be divided into an infinite number of equal and non-overlapping parts, which is true only if l is itself infinitely long. It might seem, then, that we ought to rest content with taking $\Re$ as the negation of an equivalence relation and not seek for any further extension.

There is, however, a further extension we can make, and that is to adopt the thesis also where $\Re$ is a relation *acceptable for counting the* Φ's, in the sense defined on p. 15, i.e. where $\Re$ is irreflexive and symmetrical and where in addition

$(\forall\psi)(\psi \; *\text{-}restricts_{\Re}\Phi \rightarrow$

$(\forall n)(\forall\beta)(\psi\beta \; \& \; (\exists n_{\Re}'\alpha)(\psi\alpha) \rightarrow (\exists n_{\Re}\alpha)(\psi\alpha \; \& \; \alpha\Re\beta))).$

To make *this* extension would be equivalent to adopting the axiom of choice that we have been discussing, i.e. the Principle of Dependent Choices. Since a demonstration of the equivalence would be a little tedious and since I have already expressed some doubt about the usefulness of the concept 'acceptable for counting the Φ's' (as defined), I shall here only indicate the connection between the two theses by giving an illustration of the sort of use I shall later be making of this version of the axiom of choice. In order to avoid trespassing on the material of the next chapter (§4), I here give the example in terms of classes.

Suppose, then, that we have an infinite class α, some members of which bear to one another the irreflexive and symmetrical relation R. Our question is whether α contains a denumerable subclass in which *every* member bears R to every other. Using our principle for constructing progressions we shall be able to show that this is the case provided that for every *finite* subclass of α in which every member bears R to every other there exists a further member of α, not in that subclass, which bears R to every member of the subclass. (Of course it is quite obvious to intuition that in these circumstances α must contain the denumerable subclass we require, but to *prove* the point an appeal to our principle is needed.)

We may conveniently start by abbreviating

$$Disc(\beta) \quad \text{for} \quad (\forall xy)(x{\in}\beta \ \& \ y{\in}\beta \ \& \ x{\neq}y \rightarrow xRy).$$

Now we have obviously

(i) $(\exists\beta)(\beta \subseteq \alpha \ \& \ Disc(\beta) \ \& \ (1x)(x{\in}\beta))$.

If in addition

$(\forall n)(\forall\beta)(\beta \subseteq \alpha \ \& \ Disc(\beta) \ \& \ (nx)(x{\in}\beta) \rightarrow$
$(\exists y)(y{\in}\alpha \ \& \ (\forall x)(x{\in}\beta{\rightarrow}xRy)))$

then also

(ii) $(\forall n)(\forall\beta)(\beta \subseteq \alpha \ \& \ Disc(\beta) \ \& \ (nx)(x{\in}\beta) \rightarrow$
$(\exists\gamma)((\exists y)(\gamma = \beta \cup \{y\}) \ \& \ \gamma \subseteq \alpha \ \& \ Disc(\gamma) \ \& \ (n'x)(x{\in}\gamma)))$.

However, (i) and (ii) constitute the premises for an application of our principle with

$\beta \subseteq \alpha \ \& \ Disc(\beta) \ \& \ (nx)(x{\in}\beta)$ in place of $\mathscr{A}(n, \beta)$

$(\exists y)(\gamma = \beta \cup \{y\})$ in place of $\beta\mathfrak{R}\gamma$.

We therefore deduce the existence of a function $\mathbf{f}$ from the natural numbers to the subclasses of α such that

$(\forall n)(\mathbf{f}(n) \subseteq \alpha \ \& \ Disc(\mathbf{f}(n)) \ \& \ (nx)(x{\in}\mathbf{f}(n)))$
$(\forall n)(\exists y)(\mathbf{f}(n') = \mathbf{f}(n) \cup \{y\})$.

We now define a new function $\mathbf{g}$ from the natural numbers to the members of the subclasses of this progression by putting

$$\mathbf{g}(1) = (\imath x)(x \in \mathbf{f}(1))$$

$$\mathbf{g}(n') = (\imath x)(x \in \mathbf{f}(n') \ \& \ \sim x \in \mathbf{f}(n)).$$

It is then perfectly straightforward to show as desired that

$$(\forall nm)(n \neq m \rightarrow \mathbf{g}(n) R \mathbf{g}(m)) \ \& \ (\forall n)(\mathbf{g}(n) \in \alpha)$$

which is to say

$$Disc \{x : (\exists n)(x = \mathbf{g}(n))\} \ \& \ \{x : (\exists n)(x = \mathbf{g}(n))\} \subseteq \alpha$$

and therefore

$$(\exists \infty_{\mathfrak{R}} x)(x \in \alpha)$$

This is the kind of way our principle will be applied, namely to provide a sufficient condition for the existence of a progression in which each term bears R to every other, where the relation R will be what I shall call a *discreteness* relation. That is to say, it will be used to provide a sufficient condition for applying the cardinal 'denumerably many' with a discreteness relation as counting relation. (However, we shall not be explicitly concerned with classes.)

Note on Eliminating Variables for Functions

In several places during the course of this chapter I have used variables $\mathbf{f}$, $\mathbf{g}$, ... for arbitrary functions fron natural numbers to other things. This use will recur in the next chapter (pp. 135–7), where for convenience I simply identify a progression with such a function, using $\mathbf{f}(n)$ to denote the nth term of the progression in question. My reasoning is given informally, but it could easily be formalized in such a way as to depend on three principles governing these variables for functions, namely an axiom schema[41]

FUNC 1: $(\forall n)(1\alpha)(\mathscr{A}(n, \alpha)) \rightarrow (\exists \mathbf{f})(\forall n)(\mathscr{A}(n, \mathbf{f}(n))$

and two axioms

FUNC 2: $(\forall \mathbf{f})(\forall n)(\exists \alpha)(\alpha = \mathbf{f}(n))$

FUNC 3: $(\forall \mathbf{f})(\forall nm)(n = m \rightarrow \mathbf{f}(n) = \mathbf{f}(m)).$

---

[41] Recall that '$(1\alpha)(-\alpha-)$' is the strong numerical quantifier 'there is exactly one α such that $-\alpha-$'.

Obviously the usual analysis of functions in terms of many–one dyadic relations between numbers and other things would justify these axioms, but there is a more economical analysis available which is the subject of this note. All the progressions we shall henceforth be concerned with are in fact *ordered* progressions, i.e. contain no repetitions, and in this special case we can eliminate our functional variables in favour of dyadic relations between the items in the progression. This is some advantage, because these dyadic relations are homogeneous in type and mixed-type relations between numbers and other things do not have to be invoked.

For technical convenience I prefer to work with reflexive and anti-symmetric relations rather than irreflexive and asymmetrical ones, so I now redefine[42]

$$\Re \; orders \; \{\alpha : \Phi\alpha\}$$

for

$$(\forall\alpha\beta\gamma)(\Phi\alpha \; \& \; \Phi\beta \; \& \; \Phi\gamma \to (\alpha \, \Re\beta \; \& \; \beta \, \Re\gamma \to \alpha \, \Re\gamma))$$
$$\& \; (\forall\alpha\beta)(\Phi\alpha \; \& \; \Phi\beta \to (\alpha \, \Re\beta \; \& \; \beta \, \Re\alpha \to \alpha = \beta))$$
$$\& \; (\forall\alpha\beta)(\Phi\alpha \; \& \; \Phi\beta \to (\alpha \, \Re\beta \lor \beta \, \Re\alpha)).$$

Where $\Re$ orders $\{\alpha : \Phi\alpha\}$ in this sense, we can evidently define the nth Φ in the ordering by putting

$$n_{\Re}{}^{\text{th}}\{\alpha : \Phi\alpha\} \quad \text{for} \quad (\imath\alpha)(\Phi\alpha \; \& \; (n\beta)(\Phi\beta \; \& \; \beta \, \Re\alpha)).$$

Hence we can define what it means for an ordering to be a progression of the α such that $\Phi\alpha$ by putting

$$Prog_{\Re}\{\alpha : \Phi\alpha\}$$

for

$$\Re \; orders \; \{\alpha : \Phi\alpha\}$$
$$\& \; (\forall\alpha)(\Phi\alpha \to (\exists n)(\alpha = n_{\Re}{}^{\text{th}}\{\alpha : \Phi\alpha\}))$$
$$\& \; (\forall n)(\exists\alpha)(\alpha = n_{\Re}{}^{\text{th}}\{\alpha : \Phi\alpha\})).$$

Finally I say that a dyadic relation *is* a progression if it is a progression of its field, i.e.

$$Prog(\Re) \quad \text{for} \quad Prog_{\Re}\{\alpha : (\exists\beta)(\alpha \, \Re\beta \lor \beta \, \Re\alpha)\}$$

and I introduce a quasi-functional notation for the nth member of this progression by further abbreviating

$$\Re(n) \quad \text{for} \quad n_{\Re}{}^{\text{th}}\{\alpha : (\exists\beta)(\alpha \, \Re\beta \lor \beta \, \Re\alpha)\}.$$

The method of eliminating functional variables is then to take it that they are a short way of generalizing over all ordering relations which are progressions in this sense. That is,

$$(\forall\mathbf{f})(-\mathbf{f}(n)-)$$

---

[42] The identity figuring in this definition should be taken for future applications in the sense defined on p. 113.

is construed as short for

$$(\forall \, \Re)(Prog(\, \Re) \rightarrow (— \, \Re(n) —)).$$

(Note that **f** *only* occurs in the context $\mathbf{f}(n)$.)

From these definitions we have as obvious consequences

$$\text{FUNC 2}: (\forall \mathbf{f})(\forall n)(\exists \alpha)(\alpha = \mathbf{f}(n))$$

$$\text{FUNC 3}: (\forall \mathbf{f})(\forall nm)(n = m \rightarrow \mathbf{f}(n) = \mathbf{f}(m))$$

and we also have a further consequence

$$\text{FUNC 4}: (\forall \mathbf{f})(\forall nm)(\, \mathbf{f}(n) = \mathbf{f}(m) \rightarrow n = m)$$

just because ordered progressions cannot contain repetitions. Hence it follows that the suggested axiom schema

$$\text{FUNC 1}: (\forall n)(1\alpha)(\mathscr{A}(n, \alpha)) \rightarrow (\exists \mathbf{f})(\forall n)(\mathscr{A}(n, \mathbf{f}(n))$$

is not forthcoming without qualification, and in its place we must content ourselves with

$$\text{FUNC 1}': (\forall n)(1\alpha)(\mathscr{A}(n, \alpha)) \, \& \, (\forall nm)(\forall \alpha)(\mathscr{A}(n, \alpha))$$
$$\& \, \mathscr{A}(m, \alpha) \rightarrow n = m) \rightarrow (\exists \mathbf{f})(\forall n)(\mathscr{A}(n, \mathbf{f}(n)).$$

This is quite easily deduced from the suggested definition of functional variables. For given the hypothesis

$$(\forall n)(1\alpha)(\mathscr{A}(n, \alpha)) \, \& \, (\forall nm)(\forall \alpha)(\mathscr{A}(n, \alpha) \, \& \, \mathscr{A}(m, \alpha) \rightarrow n = m)$$

we have only to consider the relation $\Re$ such that

$$(\forall \alpha \beta)(\alpha \, \Re \beta \leftrightarrow (\exists nm)(\mathscr{A}(n, \alpha) \, \& \, \mathscr{A}(m, \beta) \, \& \, n \leqslant m))$$

and it will then follow that

$$Prog(\, \Re) \, \& \, (\forall n)(\mathscr{A}(n, \, \Re(n))).$$

On this understanding of functional variables we shall have to make some revision in the principle of dependent choices, for if

$$(\exists \alpha)(\mathscr{A}(1, \alpha)) \, \& \, (\forall n)(\forall \alpha)(\mathscr{A}(n, \alpha) \rightarrow (\exists \beta)(\alpha \, \Re \beta \, \& \, \mathscr{A}(n', \beta)))$$

it certainly does *not* follow that there exists a progression **f** *without repetitions* such that

$$(\forall n)(\mathscr{A}(n, \mathbf{f}(n)) \, \& \, \mathbf{f}(n) \, \Re \mathbf{f}(n')).$$

The simplest modification is no doubt to restrict the principle of dependent choices by adding a suitable condition on the relation $\Re$ involved, for example that it be transitive and irreflexive. This condition is easily satisfied in my subsequent uses of the principle by an appropriate choice of $\Re$.[43]

---

[43] Namely '$x \subset y \, \& \, y \not\subset x$' for the first two applications, and $x < y$ for the third. The condition mentioned is stronger than necessary. A condition that is both necessary and sufficient is that when $\Re$ is confined to $\{\alpha : (\exists n)(\mathscr{A}(n, \alpha))\}$ the proper ancestral of the resulting relation should be irreflexive.

My main text will continue to use the variables **f**, **ǵ**, . . . as if they represented arbitrary functions from natural numbers to other things, but it will be found that these variables could quite easily be eliminated in the manner here sketched.

2. Rational Numbers

1. The Classical Accounts

A numeral for a rational number is constructed from two numerals for natural numbers, as in '$\frac{2}{3}$', and this is clearly no accident. We do indeed recognize that a rational number is determined by an ordered pair of natural numbers; in fact we often think of the rational number n/m precisely as the number which results from taking the natural number n and dividing it by the natural number m. Now the orthodox logicist view of rational numbers, though it denies that rational numbers can ever be obtained by dividing one natural number into another, does depend upon this fact that every rational number is determined by an ordered pair of natural numbers, and conversely that every ordered pair of natural numbers determines some rational number. Since several different pairs of natural numbers will determine the same rational number—$\frac{2}{3}$ is the same number as $\frac{4}{6}$—we cannot of course say that a rational number just is an ordered pair of natural numbers and leave it at that, but still this is the essential idea behind the logicist approach. It would be possible, indeed, to single out one preferred pair of natural numbers to identify with each rational number, for example by restricting our attention to pairs of mutually prime natural numbers, but it is usual to treat all pairs that determine the same rational number impartially. This is done by identifying the rational number with *the class of all* ordered pairs of natural numbers that determine it in this way. Thus we have

$$\tfrac{2}{3} = \{\langle 2, 3 \rangle, \langle 4, 6 \rangle, \langle 6, 9 \rangle, \langle 8, 12 \rangle \ldots\}$$

and more generally

$$\frac{n}{m} = \{x : (\exists pq)(x = \langle p, q \rangle \ \& \ p \cdot m = q \cdot n)\}.$$

This analysis of rational numbers is not of course peculiar to Frege and Russell and their followers[1], but I shall regard it as the orthodox logicist account of the matter.

Now it is easily seen that the objections we raised against identifying *natural* numbers with sets of any sort will apply equally against this proposal to identify rational numbers with certain sets. First, we do not in fact think of rational numbers as being sets—or even ordered pairs—and we find it extremely unnatural to use the set-theoretical vocabulary of membership, inclusion, and so on in connection with them. Second, and more conclusively, we can argue as before that if rational numbers really are sets then it cannot be a wholly arbitrary matter *which* sets they are, but in the present case there is no escape from arbitrariness. The only restriction there is on which sets may be chosen to be the rational numbers is that the identification should be 'adequate for all mathematical purposes', and in particular if we are choosing sets somehow constructed from the natural numbers the object will be to ensure that the arithmetic of rational numbers is deducible, via the identification, from the arithmetic of natural numbers. Now in the present case the only feature of rational arithmetic that is given directly by the identification is the number of rational numbers (denumerably many) and their chief identity-condition:

$$\frac{n}{m} = \frac{p}{q} \leftrightarrow n \cdot q = m \cdot p.$$

(We shall require further definitions for the addition and multiplication of rational numbers, and so on.) It is easily verified that these same features will be equally forthcoming from any identification of the form

$$\frac{n}{m} = \{x : (\exists pq)(x = \langle p, q \rangle \ \& \ p \cdot m = q \cdot n \ \& \ (\ldots p \ldots q \ldots))\}$$

so long as we can prove of the extra clause $(\ldots p \ldots q \ldots)$ that

$$(\exists pq)(p \cdot m = q \cdot n \ \& \ (\ldots p \ldots q \ldots))$$

Of course there are many such extra clauses for which this

[1] So far as I know the analysis first appears in Frege 1884 (§104). But Peano 1901 (pp. 54f) is often cited as a source, and this may well be independent of Frege.

can be proved. One, already hinted at, is

p and q are mutually prime

another, evidently arbitrary, is

$$p > 99$$

and so on. However, all these different choices of the sets to take as identical to the rational numbers are equally 'adequate for all mathematical purposes', and consequently none of them is philosophically adequate.

In the case of the natural numbers our response to this objection was to forgo any identification of natural numbers with sets or other objects and instead to turn our attention to the analysis of whole statements apparently about natural numbers. The same response can be pursued here, and indeed it yields a very simple account of rational arithmetic. First, we may straightforwardly define the whole statements:

$$\frac{n}{m} = \frac{p}{q} \quad \text{for} \quad n \cdot q = m \cdot p$$

$$\frac{n}{m} < \frac{p}{q} \quad \text{for} \quad n \cdot q < m \cdot p$$

and in conjunction with the standard definitions

$$\frac{n}{m} + \frac{p}{q} \quad \text{for} \quad \frac{n \cdot q + m \cdot p}{m \cdot q}$$

$$\frac{n}{m} \cdot \frac{p}{q} \quad \text{for} \quad \frac{n \cdot p}{m \cdot q}$$

this will allow us to translate into the notation of elementary arithmetic any statement of rational arithmetic which can be expressed using only schematic letters and variables (and constants) for natural numbers. When next we consider statements using variables for rational numbers, say $r, s, t, \ldots$, the relevant definitions may evidently be given as

$$(\forall r)(\ldots r \ldots) \quad \text{for} \quad (\forall nm)\left(\ldots \frac{n}{m} \ldots\right)$$

$$(\exists r)(\ldots r \ldots) \quad \text{for} \quad (\exists nm)\left(\ldots \frac{n}{m} \ldots\right)$$

and the translation rules are then complete. For example the true statement of rational arithmetic

$$(\forall rs)(\exists t)(r \cdot t = s)$$

is successively translated into

$$(\forall nmpq)(\exists kl)\left(\frac{n}{m} \cdot \frac{k}{l} = \frac{p}{q}\right)$$

$$(\forall nmpq)(\exists kl)\left(\frac{n \cdot k}{m \cdot l} = \frac{p}{q}\right)$$

$$(\forall nmpq)(\exists kl)(n \cdot k \cdot q = m \cdot l \cdot p)$$

which last is a truth of elementary arithmetic. The whole of the arithmetic of rational numbers can easily be constructed in this way from elementary arithmetic[2], and no difficulty arises from our refusal to treat n/m as standing for anything by itself.

Nevertheless the construction is still, I think, philosophically inadequate. Essentially the objection to it is that it provides us with no real understanding of the concept of a rational number; it tells us how parts of elementary arithmetic can be rewritten so as to look as if they yield statements about some new numbers, different from the natural numbers, but it gives us no indication of the point (if any) of such a rewriting. Certainly it does very little to persuade us that there is any point in construing $\frac{2}{3}$ as an expression for one single number. More precisely, I think that this objection could be pressed by asking for an explanation of why in rational arithmetic we naturally use the same signs $<$, $+$, and $\cdot$ as are already used in elementary arithmetic, and why we choose to define these signs in rational arithmetic in the way that we do. There is no doubt that we do think it appropriate to speak of the *addition* and *multiplication* of rational numbers, just as we speak of the addition and multiplication of natural numbers, so at the very least we must think that there is some analogy between the uses of $+$ and $\cdot$ in these two different contexts. But what is this analogy? Why, for example, do we not define

$$\frac{n}{m} + \frac{p}{q} \quad \text{for} \quad \frac{n \cdot p}{m \cdot q}$$

[2] Compare Quine, 1970, pp. 75–6.

$$\frac{n}{m} \cdot \frac{p}{q} \quad \text{for} \quad \frac{n \cdot q + m \cdot p}{m \cdot q} \; ?$$

One answer that might be suggested is that the distributive law which holds for natural numbers, namely

$$n \cdot (p + q) = n \cdot p + n \cdot q$$

would no longer hold for rational numbers if these alternative definitions were adopted. Generally it might be said that we choose the definitions we do choose just because these definitions preserve as many as possible of the formal properties of $<$, $+$, and $\cdot$ between natural numbers. However, this can easily be shown to be an inadequate answer to our problem, because the same could be said of infinitely many other ways of defining $<$, $+$, and $\cdot$ between rational numbers. To see this we have only to consider any permutation of the rational numbers, i.e. any function f from rational numbers to rational numbers which also has a unique inverse f^{-1}, and in terms of this to define new signs '$\oslash$', '$\oplus$', '$\otimes$' applicable to any rational numbers r and s by putting

$$r \oslash s \quad \text{for} \quad f(r) < f(s)$$

$$r \oplus s \quad \text{for} \quad f^{-1}(f(r) + f(s))$$

$$r \otimes s \quad \text{for} \quad f^{-1}(f(r) \cdot f(s)).$$

It is obvious that the formal properties of $\oslash$, $\oplus$, and $\otimes$ must then correspond exactly to the formal properties of $<$, $+$, and $\cdot$ between rational numbers, and so according to the present suggestion each must have an equally good claim to be taken as corresponding to $<$, $+$, and $\cdot$ between natural numbers. But clearly $\oslash$ and $<$, $\oplus$ and $+$, and $\otimes$ and $\cdot$ need by no means coincide in their application to rational numbers. Nor is the difficulty avoided if we further stipulate (and why should we?) that the rational number $n/1$ should behave in all ways like the natural number n, so that

$$\frac{n}{1} < \frac{m}{1} \leftrightarrow n < m$$

$$\frac{n}{1} + \frac{m}{1} = \frac{n + m}{1}$$

$$\frac{n}{1} \cdot \frac{m}{1} = \frac{n \cdot m}{1}$$

for in order to secure this result we have only to stipulate that our permutation f should obey the condition

$$f\left(\frac{n}{1}\right) = \frac{n}{1}$$

which still leaves us with any number of different permutations[3]. Generally, if the operations of addition, multiplication, and so on which are applicable to rational numbers are taken to be distinct from the corresponding operations on natural numbers, then the mere aim of preserving formal properties will not determine how these operations on rational numbers are to be understood. The orthodox logicist account cannot, I think, offer any other explanation of why these operations are understood as they are, and that seems to be a fatal defect.

It is worth noticing, however, that this defect does not arise if we take a different (but equally traditional) view about the relation between the system of rational numbers and the system of natural numbers, namely that the rational numbers are postulated to exist in order to fill a gap in the natural numbers. Any two natural numbers can of course be added, but the inverse operation of subtraction cannot always be performed— there is no such thing as the natural number which results upon subtracting n from m in cases where $n > m$—and negative numbers are often thought of as specifically introduced in order to fill this gap. Similarly, though any two natural numbers may be multiplied, the inverse operation of division cannot always be performed if we remain within the natural numbers, so rational numbers may be thought of as *added* to the system of natural numbers in order to ensure that this operation of division is always performable. The idea between this approach, then, is simply to postulate that there does exist one and only one

---

[3] For example the function f so defined that
$$f(r) = (2 \cdot r - n) + 1 \quad \text{if} \quad n < r + 1 \leqslant n + \tfrac{1}{3}$$
$$f(r) = \tfrac{1}{2} \cdot (r + n) \quad \text{if} \quad n + \tfrac{1}{3} < r + 1 \leqslant n + 1$$
is monotonically increasing and satisfies the condition $f(n/1) = n/1$. However, $\oplus$ and $\otimes$ defined in terms of it do not in general coincide with the ordinary $+$ and $\cdot$.

number which is the result of dividing any natural number by any other, and the new numbers thus postulated we shall call rational numbers.

In more detail, we might conveniently define

$$x \text{ is a rational number} \quad \text{for} \quad (\exists nm)(m \cdot x = n)$$

and henceforth (for brevity) we shall confine our variables r, s, ... to rational numbers according to this definition. (Notice, incidentally, that according to this definition every natural number is also a rational number, i.e. we have

$$(\forall n)(n \text{ is a rational number})$$

i.e. $(\forall n)(\exists x)(x \text{ is a rational number} \ \& \ x = n)$

i.e. $(\forall n)(\exists r)(r = n)$.   (A)

Now we may postulate the existence and uniqueness of rational numbers by laying down two axioms

$$(\forall nm)(\exists r)(m \cdot r = n) \qquad\qquad\qquad \text{Ex(istence)}$$

$$(\forall nm)(\forall rs)(m \cdot r = n \ \& \ m \cdot s = n \rightarrow r = s). \ \text{Un(iqueness)}$$

In these postulates we are of course already employing the idea of multiplying a rational number by a natural number, since this is inevitably involved in the idea that rational numbers result from dividing one natural number by another. The point perhaps will be brought out more clearly if we employ the Russellian theory of definite descriptions to introduce the usual notation for rationals:

$$\frac{n}{m} \quad \text{for} \quad (\imath r)(m \cdot r = n).$$

The postulates then assure us at once that

$$(\forall nm)(\exists r)\left(r = \frac{n}{m} \right) \qquad\qquad\qquad (B)$$

(so that scope-indicators may be ignored), while conversely the definition of 'rational number' tells us that

$$(\forall r)(\exists nm)\left(r = \frac{n}{m} \right).$$

Further, the salient fact about n/m, given immediately by its definition, is

$$m \cdot \frac{n}{m} = n. \tag{C}$$

Therefore we are already stipulating, in effect, what is to be the result of multiplying the rational number n/m by the natural number m.

This does not itself tell us how multiplication of two arbitrary rational numbers is to be construed, but it turns out that there is only one way of construing this if we are to preserve the commutativity and associativity of multiplication. If we assume

$$r \cdot s = s \cdot r, \quad r \cdot (s \cdot t) = (r \cdot s) \cdot t \tag{Mult}$$

we may then argue

(i) $(m \cdot q) \cdot \left(\frac{n}{m} \cdot \frac{p}{q} \right) = \left(m \cdot \frac{n}{m} \right) \cdot \left(q \cdot \frac{p}{q} \right)$ by Mult, A, and B

$\therefore$ (ii) $(m \cdot q) \cdot \left(\frac{n}{m} \cdot \frac{p}{q} \right) = n \cdot p$ by (i) and C

$\therefore$ (iii) $\qquad \frac{n}{m} \cdot \frac{p}{q} = \frac{n \cdot p}{m \cdot q}$ by (ii), C, and Un.

Similarly there will be only one way of construing the addition of two arbitrary rationals if we postulate that the distributive law is still to hold. That is, if we assume

$$r \cdot (s + t) = r \cdot s + r \cdot t \tag{Add}$$

we may then argue

(i) $(m \cdot q) \cdot \left(\frac{n}{m} + \frac{p}{q} \right) = (m \cdot q) \cdot \frac{n}{m} + (m \cdot q) \cdot \frac{p}{q}$ by Add, A, and B

$\therefore$ (ii) $(m \cdot q) \cdot \left(\frac{n}{m} + \frac{p}{q} \right) = n \cdot q + m \cdot p$ by (i), Mult, and C

$\therefore$ (iii) $\qquad \frac{n}{m} + \frac{p}{q} = \frac{n \cdot q + m \cdot p}{m \cdot q}$ by (ii), C, and Un.

(Similar results also hold for $<$.) The most evident defect of the logicist approach, then, will disappear if we do not try to present rational numbers as logical constructions from natural numbers, but instead simply postulate them as entities needed to complete the system of natural numbers in a certain way. What makes the crucial difference between the two approaches is that the postulation demands that natural numbers and rational numbers mesh together to form one coherent system (so that n and $n/1$ are the *same* number), while in the orthodox logicist view natural and rational numbers do not mix with one another (so that n and $n/1$ are *distinct* but in some way *corresponding* numbers).

Of course the second approach has a defect of its own: how do we know that the postulates required for the second approach are indeed true? By mapping the second approach onto the first we can of course easily show that these postulates are *consistent* (if elementary arithmetic is consistent), but it may well be said, as Brouwer insisted against Hilbert[4], that consistency is no guarantee of truth, even in mathematics. However, this point requires some elaboration. Certainly there do seem to be some areas of mathematics where consistency is all that we can reasonably require, and where the question of truth does not really seem apposite. Much of what is known as 'abstract algebra'—for example the theory of groups—seems properly described in this way, and this is because there does not appear to be any antecedent understanding of the relevant concept of a *group* which would make it reasonable to ask whether the usual axioms of group theory do succeed in capturing our pre-axiomatic understanding of this concept. Rather, the axioms themselves determine what is meant by a group, and hence the question of the *truth* of these axioms simply does not arise, though no doubt we should modify them in some way if they turned out to be inconsistent, and lose interest in them if they had no known models. From the point of view of the pure mathematician the same might be said of the various 'geometries' now available; no doubt geometry began as a study of the spatial properties of actually existing things, but now a mathematician will treat the traditional axioms as simply *defining* a 'Euclidean space' and will leave to others the question of whether our space is indeed like that. So again the question of

---

[4] See for example Brouwer's *Intuitionistic reflections on formalism* in van Heijenoort (1967), pp. 490–2.

whether the axioms are as they stand *true* seems to make very little sense, though the physicist can certainly ask whether they can be given an interpretation which makes them true of actual space, or of anything else. Generally, it does seem correct to regard several branches of mathematics as concerned simply to study the consequences of certain axioms which are not themselves supposed to be true in any substantial sense[5], though they may perhaps have interesting *models* or other useful *applications*.

Some formalists[6] give this account even of the theory of natural numbers. On this account, all that can be said of an axiom of natural number theory, such as 'no two natural numbers have the same successor', is that the presence of this axiom makes for an aesthetically satisfying system, and one which has more fruitful applications than would result from modifying, dropping, or denying it; since the axiom is part of the definition of 'natural number', one cannot sensibly raise the question of whether or not it is actually true of the natural numbers. However, if the concept of a natural number is thus wholly given by the axiomatizations of number theory, it would seem to follow that before such axiomatizations were produced (roughly a century ago[7]) there simply was no such thing as the concept of a natural number. At any rate it certainly does follow that the concept of a natural number could not precede the implicit recognition of sufficient laws of number theory to yield our present axiomatizations, including for example the principle of mathematical induction which was not explicitly formulated until the sixteenth century[8]. This consequence is surely preposterous. A modified type of formalism can avoid it, as I shall show later (pp. 293–7), but on the face of it the orthodox formalist account of the natural numbers must have this consequence, and so must be untenable. However, more can be said in its favour when we turn to rational numbers. As we have seen, we can supply a good formalist motive for being interested in the system of rational numbers, namely that it is aesthetically

---

[5] Of course the axioms of Euclidean geometry are by definition true of Euclidean space, and in this trivial sense they may be said to be true.

[6] For example Curry 1951.

[7] Dedekind's *Was sind und was sollen die Zahlen?* was published in 1888.

[8] See for example Kline 1972, p. 262.

satisfying to have a system of numbers in which division, as well as multiplication, is always performable. So long as we think of this system as obtained by postulating an extension of the system of natural numbers, all characteristics of the usual system of rational numbers are then readily accounted for and the postulation is easily seen to preserve consistency. Consequently in this case the formalist can give an explanation of why we adopt those axioms that we do adopt without having to admit that we have any other route to the concept of a rational number.

However, even though we have not yet forced the formalist to admit this, it seems to me that it should anyway be admitted. For example, it seems clear that we can learn to use *proper fractions* in contexts such as 'half an apple' or 'three-quarters of an hour' before we have much grasp of the formal laws that govern them, and certainly before we extend the notion to include 'improper' fractions. (How do you take three halves of *an* apple?) Until this extension is made, however, it can hardly be maintained that our understanding of fractions rests on a desire to ensure the universal performability of division, and equally it can hardly be maintained that the concept of a rational number is quite divorced from the concept of a fraction. Again it is noteworthy that the classical Greek mathematicians had a lot to say about *ratios* and *proportions* though they never recognized the existence of rational (or irrational) *numbers*, so the two concepts are not absolutely interchangeable with one another. Yet it seems clear that there is a very close connection between, say, the ratio of 3 to 2 and the rational number 3/2. What exactly this connection is we shall shortly investigate, but these brief remarks should be enough, I think, to suggest that the formalist account has left out something that is very important. Our understanding of fractions and ratios is largely independent of what the formalist sees as the main motivation for the theory of rational numbers, and so we may quite sensibly ask whether there is any reason to suppose that his postulates are true of fractions and ratios as we understand them. Further, it seems to me that the main motivation for rational arithmetic is in fact its bearing on our use of fractions and ratios and other 'applications' I shall introduce shortly, and in that case the formalist's failure to answer our question is evidently a serious gap in his

account. In order to fill the gap I shall now develop an alternative approach, in which the arithmetical theory of rational
numbers is seen much more as a theory designed to generalize
and explain our pre-arithmetical understanding of fractions,
ratios, and so forth.

In sum, both of the accounts of rational numbers that I have
called 'classical' are accounts which rely upon the relation of the
theory of rational numbers to the theory of natural numbers,
and largely ignore its relation to other areas. The orthodox
logicist account presents rational numbers as logical constructions from natural numbers, but against this I have argued that
there are features of the theory of rational numbers which such
an account cannot explain, for example why addition and multiplication are defined for rational numbers in the way that they
are defined. The formalist account presents rational numbers as
postulated to 'complete' the theory of natural numbers in a
certain way, and in so doing it avoids the previous objection.
However, in general a formalist account is acceptable only
where we are prepared to say that we have no independent
grasp of the crucial concepts of the formalized theory, and in
the case of rational numbers this condition seems not to be
satisfied. Indeed a formalist account of *natural* numbers seems
absurd just because it is clear that (logical) facts about number-
ing and counting dictate at least some of the axioms of the
theory of natural numbers, and in fact I believe—as I tried to
show in the first volume of this work—that they dictate all the
axioms. In the light of the last chapter we can also add that
these axioms seem to be equally dictated by other, equally
logical, facts about powers of relations and about ordinal adjectives. In Chapter 4 I shall consider what is to be done about this
'overdetermination' of the theory of natural numbers, but we
can surely say already that to ignore *all* these routes to the
concept of a natural number must be to leave out the heart of
the matter. In this chapter I shall pursue the view that the
formalist's basic axioms and definitions for the theory of rational
numbers are again dictated by what he would call the 'applications' of that theory, and that our grasp of the concept of a
rational number again relies on our grasp of the 'uses' or
'applications' of these numbers. Then in the next chapter I shall
try to extend the argument to cover irrational numbers. Such

an approach has more initial plausibility where the applications of a branch of arithmetic—the concepts which we now regard as somehow involving the relevant sort of number—were actually in use before the formal theory of such numbers was elaborated, and this surely was the situation with natural numbers, with rational numbers, and with real numbers. I would not say that the appeal to history settles the question of how contemporary mathematics is to be understood, but it does add some weight to the suggestion and is enough to show that the idea is worth pursuing.

Let us turn, then, to consider the applications of rational numbers, and how those applications bear upon the arithmetical theory of rational numbers.

2. Measurement

It is a common view that numbers are used chiefly in counting and in measuring: in counting we use only the natural numbers, while in measuring we use also the rational (and perhaps the irrational) numbers. Therefore when we seek an application for the rational numbers it is natural to begin by considering the topic of measurement. One interesting thing that emerges from the contemporary accounts of measurement is that they are very little concerned to describe any application of *numbers*, but at most of *numerals*. This feature is due, I think, to an excessive concentration on the value of measurement to the experimental scientist. Let us first explore this approach.

On one of the most hospitable definitions of measurement it is 'the assigning of numerals to objects or events according to rules—*any* rules'[9], but this is clearly too hospitable to be at all close to our common view. When the members of a football team are each assigned a numeral as a label, we do not commonly think that anything has thereby been *measured*. More commonly we think of measurement as applying to *quantities*, such as length, weight, size, temperature, speed, and so on, which are capable of different *degrees*. Each of these quantities is associated with a weak *ordering* relation, such as 'longer than', 'heavier than', 'larger than', and a corresponding *equivalence*

[9] Stevens 1946.

relation 'as long as', 'as heavy as', 'as large as'. These relations give rise to the distinct degrees, for clearly different things are said to have a greater, equal, or lesser degree of some quantity φ-ness according as they are more φ than, as φ as, or less φ than one another. It is not easy to characterize just what is meant by a quantity and its degrees, but the concept is no doubt sufficiently familiar for us to leave its analysis till later[10]. Anyway, let us henceforth confine our attention to the measurement of such quantities, for that is surely the primary application of measurement.

Now one clear use of measurement is that it enables us to describe the 'quantitative' features of phenomena very much more precisely than we could otherwise do, and this is of value to the scientist since it is largely these features which turn out to be connected with one another in the systematic ways which it is his aim to discover. For example, our scientist may well suspect that there is a systematic connection between the *length* of a copper bar and its *temperature*, between the *volume* of a mass of air and the *pressure* it exerts on its container, between the *weight* of a falling object and the *depth* of the crater it makes, and so on. However, before he can give any precise formulation of this connection, he must first have some effective way of describing the different lengths, temperatures, and volumes that an object may have; he must have a way of labelling the various degrees of these quantities, and therefore he needs a set of labels and an agreed method of assigning them.

In a few cases the number of distinct degrees he wishes to consider may be small enough for him to assign them labels one by one (perhaps ostensively), as for example with Moh's 10-degree scale of hardness for minerals. However, this is rare. In most cases the quantities he is interested in will not appear to have either a lowest degree or a highest degree, and it will usually seem that between any two distinct degrees there is yet a further degree, so that the different degrees will apparently fall into a dense order with no first or last member. It follows that infinitely many labels will be needed if we are to be sure of discriminating between any two degrees, and it will be natural to choose a set of labels which is itself easily arranged in a dense order with no first or last member so that the order of the

<hr/>

[10] See §6 of the present chapter.

different degrees can be mirrored by an easily discerned ordering of the labels. These considerations suggest that the standard
numerals for the (positive) rational numbers will be suitable
for use as labels, since they do of course have the required type
of order, but so far we have had no special reason to employ
just these labels rather than any others. Obviously one possible
alternative would be to use the numerals for the positive
rationals less than some given rational, or we could invent a
new set of labels for the purpose. To illustrate, we might use as
a new set of measuring labels any finite string of 0's and 1's
that ends with a 1, and we might define an order for them by
considering the digits that occur in their first place, their second
place, their third place, and so on. Thus the order of any two
labels is determined by the first place (counting from the left)
in which they differ, and a label with no digit in that place is
earlier than a label with a 0 in that place, which in turn is earlier
than a label with a 1 in that place. (For example the labels of
two digits or less, in their correct order, are

$$01, 1, 11;$$

the labels of three digits or less in their correct order are

$$001, 01, 011, 1, 101, 11, 111;$$

the labels of four digits or less in their correct order are

$$0001, 001, 0011, 01, 0101, 011, 0111, 1, 1001, 101, 1011,$$

$$11, 1101, 111, 1111;$$

and so on. It is easily seen that the ordering is indeed a dense
ordering with no first or last member.) I think it is helpful to
keep in mind such alternatives as these just because there is no
one natural way of correlating these new labels one-to-one with
the ordinary rational numbers.

   Having settled upon the labels he intends to use, our scientist
now has to set up rules for assigning them to the different
degrees of some quantity, and for illustration let us take the
quantity length. There seems to be no avoiding the fact that to
begin with at least one label will have to be assigned to a particular length directly (and we may call this the 'unit' length),

but there are any number of different sorts of rules one might use for assigning the rest. If we suppose that 1 (i.e. 1/1) is chosen as our unit label, our usual scale of length would result if we next selected a simple progression starting from the unit label, namely

$$1/1,\ 2/1,\ 3/1,\ 4/1,\ \dots$$

and gave a recursive stipulation: if l_1 is the unit length, and l_2 the length assigned to a numeral $n/1$, then the length to be assigned to the succeeding numeral $(n+1)/1$ is the length of a composite rod formed by placing end to end in a straight line two rods of lengths l_1 and l_2. Having thus assigned all our numerals of the form $n/1$, we then add a further rule assigning n/m in terms of $n/1$, which of course is easily done.

It should be clear that we can very easily vary our stipulations so as to produce different but equally effective scales. To illustrate, here is a simple procedure for measuring lengths by assigning to them the numerals for the proper fractions, i.e. the rational numbers less than 1. First we select some standard length arbitrarily. Now to find the length to be assigned to the numeral n/m ($n < m$) we proceed as follows. First draw lines

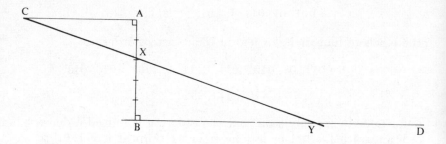

AB and AC each of the standard length and meeting at right-angles. Next draw BD of arbitrary length parallel to AC. Now divide AB into m equal parts, and, counting from B, mark the nth of the $m-1$ dividing points. Call this point X. Join CX and produce it to meet BD at Y. Then the numeral n/m is to be assigned to the length of BY. It is easy to check that as a result of these stipulations the order of the labelled lengths is indeed mirrored by the standard order of the numerals that label them. Further, every length comes to lie between two labelled lengths, and any two distinct lengths are separated by a labelled length, so we may certainly use this system to classify and discriminate between lengths just as finely as with the usual system[11]. In fact this unusual scale has been chosen so as to have a very simple relationship to the usual one, for, supposing that our standard length is assigned the numeral 1 in the usual method of measuring, a length which is assigned the numeral n/m on this unusual scale is assigned the numeral $n/(m-n)$ on the usual scale.

As a further illustration, here is a less orthodox measuring procedure making use of some special properties of the unorthodox measuring labels introduced on p. 91. This time we shall start with a combined progression and reverse progression meeting in the unit label:

$$\ldots, 0001, 001, 01, 1, 11, 111, 1111, \ldots .$$

Choosing a length for our unit label 1, we assign all the above labels by stipulating that if l_1 is the length of the side of a square whose diagonal is of length l_2 then l_1 is assigned to one of the above labels if and only if l_2 is assigned to its successor in this ordering. To complete the assignation we next define the

principal neighbours of a label thus: the earlier (later) principal neighbour of a label is the latest (earliest) label which is earlier (later) than it and which has fewer digits than it. Now we notice that each of the remaining labels has one earlier and one later principal neighbour, and stipulate that if the principal neighbours of a label are assigned to lengths l_1 and l_2 then the label

---

[11] Of course if (as we believe) there are incommensurable lengths, *no* system of measurement which uses only denumerably many labels will succeed in labelling *all* the lengths.

itself is to be assigned the length l_3 which is the length of the side of a square whose diagonal is also the diagonal of a rectangle with sides of lengths l_1 and l_2. The relation between the scale resulting from this stipulation and the usual one is a little

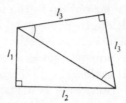

complex[12], but again the system of measurement defined has all the required features. That is, the order of the labelled lengths is mirrored by the order of their labels, every length falls between two labelled lengths, and every two distinct lengths are separated by a labelled length. Therefore (if we discount the fact that its unorthodox labels are a little unwieldy in practice) this system of measurement is as good as any other for the purpose of describing, classifying, and discriminating lengths.

This second illustration is aimed to reinforce my contention that *measurement*, as currently conceived, need have no very noticeable connection with the familiar rational *numbers*. Of course in practice we find it more convenient to employ the

---

[12] The relation can be given thus. First each measuring label is associated with an expression for a positive rational number in three steps: (i) if the label begins with a 1, then if there are n 1's to the left of the first 0 (if any) we exchange those n 1's for 2^{n-1}; (ii) the first 0 occurring (if any) is deleted; (iii) the remaining digits are treated as a positional binary notation for sums of decreasing integral powers of 2, starting from 2^{n-2} if step (i) above applies and from 2^{-1} otherwise. Thus

$$1110100101$$

becomes

$$2^2 + 2^1 + 0 + 0 + 2^{-2} + 0 + 2^{-4} \quad (=\tfrac{201}{32})$$

and

$$0001101$$

becomes

$$0 + 0 + 2^{-3} + 2^{-4} + 0 + 2^{-6} \quad (=\tfrac{21}{128}).$$

Then the relation between the unorthodox scale and the usual one is that a length which is assigned some number on the usual scale is assigned the label associated with the square of that number on the unorthodox scale.

familiar numerals in this task, rather than to invent such new and unorthodox measuring labels as I have been considering, but it should be clear that any connection between these familiar numerals and the rational numbers is so far entirely superfluous. The only relevant connection so far is that both the measuring labels and the rational numbers fall into a dense order with no first or last member. This *enables* us to set up an order-preserving correlation between them if we want to, but so far we have not wanted to. (If we assume that our usual method of measuring length is somehow the 'natural' one, then a 'natural' (but partial) correlation is given by the measuring procedure itself, as in footnote 12. So far, however, we have no warrant for picking any one system of measurement as 'natural').

However, it may be replied that this point, true though it is, has after all no force towards showing that the scientist could in the end achieve his aim without any use of numbers. For we began by saying that the scientist needs to be able to label the various degrees of a quantity merely as a preliminary to his ultimate aim of formulating the systematic connections which obtain between various quantities, and perhaps the fact is that he could not formulate these connections if he did not use for his measuring labels some symbols that were essentially connected with numbers. After all, he does certainly put forward his results in the form of 'numerical laws', and surely these must have some connection with numbers?

Now it is true that the numerical laws scientists actually put forward do have some connection with numbers, but again this connection is theoretically superfluous. For instance, if we consider Boyle's law as an example, the usual statement of this law would be that for a given mass of gas maintained at constant temperature, the volume of the gas varies inversely as the pressure it exerts on its container, or in symbols $PV = k$. Now when we reflect on the information contained in this statement we see that it is just the information that if, under the conditions proposed, the pressure and volume of the gas are measured in the normal way and are assigned the numerals 'n' and 'm' respectively, then the product of the numbers n and m for which these are the numerals will always be the same (and in fact equal to the number with numeral 'k'). Essentially, then, the law states that there will always be a certain relation between the numerals '*n*' and '*m*' assigned to the pressure and volume of the gas (when these numerals are assigned in the normal way), and this relation is then described by means of a relation between the corresponding numbers *n* and *m*, namely that $n \cdot m = k$. However, that is the only way in which numbers enter into the law, and it is clear that we would not change the information conveyed if we employed some other way of describing the relevant relation between numerals. Further, there is evidently no reason why this should not be done, for the operation on numerals which reflects the multiplication of rational numbers can perfectly well be introduced directly. In fact there is no essential difficulty in specifying an operation, say composition, to be performed on our unorthodox measuring labels which would 'correspond' to multiplication in this sense: if l_1, l_2, l_3 are three lengths so related that in the usual system of measurement they are assigned numerals 'n_1', 'n_2', 'n_3', and it is an arithmetical truth that $n_1 = n_2 \cdot n_3$, then also in the unorthodox system they are assigned labels x_1, x_2, x_3 such that x_1 is the composition of x_2 and x_3. To introduce this operation it is clear that we do not *need* to invoke any correlation between the unorthodox labels and numbers[13].

My conclusion so far is that when we consider just the requirements of the experimental scientist we find that his use of measurement need have nothing to do with our familiar conception of numbers. He does need to be able to label the various degrees of the quantities he is concerned with, and for this purpose it is certainly convenient to correlate them with the rational numbers, but (i) *any* effective correlation is as good as

[13] Since it is a little tedious to introduce the full definition I here content myself with a partial characterization.

Suppose we define a successor-function **S** on all our measuring labels thus

  (i) $\mathbf{S}('1'^{\frown}x) \doteq '11'^{\frown}x$ (and $\mathbf{S}('1') = '11'$)

  (ii) $\mathbf{S}('01'^{\frown}x) = '10'^{\frown}x$ (and $\mathbf{S}('01') = '1'$)

  (iii) $\mathbf{S}('00'^{\frown}x) = '0'^{\frown}x$.

(Here $^{\frown}$ symbolizes concatenation. Thus $'1'^{\frown}x$ means the label obtained by writing '1' followed by the label x.) Then the operation $*$ of composition obeys the recursive laws

  (iv) $x * '1' = x = '1' * x$

  (v) $x * \mathbf{S}(y) = \mathbf{S}(x * y) = \mathbf{S}(x) * y$.

These suffice to determine $x * y$ in many cases, but not yet in all.

any other, and (ii) in fact any set of labels which falls naturally into a dense order (with no first or last member) will do equally well instead of the rational numbers. He also needs to state the relationships obtaining between the degrees of various quantities, and for this purpose it is again convenient to make use of the familiar relationships between rational numbers, but again this is not essential. The desired relationships *could* be introduced in ways which make no reference to numbers, and no relevant information would thereby be lost.

At this point someone might object: if operations corresponding to addition and multiplication and so on were thus introduced for our unorthodox measuring labels, and if we performed calculations with them of a similar nature to the calculations we perform with numbers, why should we not say that these labels are themselves numbers? I do not think that this would be a very useful way of talking, for the relevant analogy seems to be that these labels are tools used in measuring in much the same way as pebbles may be used in counting, but we do not think that pebbles ('counters') are themselves numbers and so we should not think that these labels ('measurers'?) are numbers either. More importantly, however, even if we did grant that these labels may be called 'numbers' without impropriety, still it is clear that we have no warrant to identify them with (any subset of) the familiar rational (or real) numbers, and it is those familiar numbers that are my main concern. For it is a feature of the familiar rational (and real) numbers that they *include* the natural numbers and that their order with respect to the natural numbers is quite definite. However, this does not hold for the unorthodox measuring labels.

This conclusion may be reinforced by a different line of thought. It would commonly be held that in any procedure for measuring length the choice of a unit length is quite arbitrary, and therefore there can be no significance in the fact that such and such a particular number is assigned to this or that particular length. However, it would also be held that—with the usual method of measuring length—the *ratios* between the numbers assigned to different lengths are significant: if the one number is, say, twice the other, then this reflects the fact that the one length is actually *twice as long* as the other. However, our discussion of measurement so far has said nothing about this aspect of measurement. In measuring a quantity all that we have required so far is that the *order* of the measuring labels reflects the *order* of the degrees labelled (and, ideally, that all degrees fall within the measuring system). If we also wish the *ratios* of the degrees to be reflected in the *ratios* of the numbers assigned to them, clearly we must impose some additional requirements.

But then the question arises whether it really does make sense to speak of the *ratios* of the degrees of a quantity, and I think this is in fact the heart of the question whether numbers do apply to quantities in any non-arbitrary way. In some cases it seems fairly clear that our actual measuring procedures do *only* reflect the order of the degrees of a quantity, and contain no information about ratios. As an example we might cite the case of temperature. Here it is evident at once that measurements made on the Centigrade or Fahrenheit scales yield no information about ratios, just because each scale is clearly as good as the other and each would yield quite different results. However, it may be argued that the so-called 'Absolute' scale of temperature is superior to either of these scales, and one might suggest that it would yield the required basis, for why should we not say that if x is at n °A and y at $2n°$ A then y is *twice as hot* as x? Well, first it seems relevant to notice that we do not in fact say this, e.g. we do not in fact regard $+273°$ C as being *twice as hot* as $0°$ C, and second that though we might perhaps come to stipulate that the expression is to be understood in this way such stipulation would still be essentially arbitrary. For it certainly is not the case that our current 'Absolute' scale of temperature is the only scale which could be used for this purpose.

The current 'Absolute' scale has two clear advantages over, say, the Fahrenheit and Centigrade scales. First, it has a non-arbitrary zero, in that we have reason to believe that no body can have a temperature less than $0°$ A while all temperatures greater than $0°$ A may perfectly well be exemplified; and second the scale is an 'absolute' one in the sense that it is defined without reference to the heat-related properties of any particular substance, whereas the Centigrade scale, for instance, would generally be regarded as defined specifically with reference to the expansion of *mercury*. But there is no doubt that these two advantages may equally well be possessed by other scales of temperature. For instance, just as the current 'Absolute' scale is defined by reference to the efficiency of a perfectly reversible heat engine, so we could equally define a new 'absolute' scale by reference to the radiation of a perfectly black body (the perfectly black body being just as much a theoretical ideal as the perfectly reversible heat engine). However, on the most natural way of setting up such a scale one would find that n degrees on the new scale would correspond to n^4 degrees on the old[14], and so temperature ratios would certainly not be preserved. It seems quite clear that no good reason could be given for preferring the one scale to the other. Since we have a determinate mathematical transformation from one scale to the other, any physical laws which can be expressed with respect to the one scale can be expressed equally well with respect to the other, so each of them is perfectly adequate for the purposes of the scientist. There is, it would seem, only a historical reason for choosing the one we do choose. (Hence I cannot attach any sense to the contention that the Centigrade and Fahrenheit scales, though not 'ratio' scales, are at least 'interval' scales[15]. It seems that this *should* mean that equal intervals between the numbers assigned to temperatures on these scales reflect what actually are equal intervals of temperature, so that for example a rise in temperature from 0 °C to 10 °C really is as great a rise in temperature as that from 90 °C to 100 °C. However, the only backing to this contention seems to be that the two scales agree with *one another* in this matter (and, roughly, with the Absolute scale), not that they somehow agree with reality.)

Anyway, in some cases I think it would be generally admitted that we have no way of determining the 'true' ratios between the different degrees of a quantity, and though we may perhaps stipulate some conventional way of understanding these ratios, the stipulation, being essentially arbitrary, is of no interest. So far as the purposes of experimental science are concerned, it would not in the least matter if this were the case with every quantity, but the common sense view is that in some cases we can do very much better. The common sense view is surely that when we speak of one object as being three times as large as

---

[14] For details see Ellis 1966, p. 95 (from whom I take this example).
[15] See e.g. Stevens 1946, p. 146.

another, or a third as heavy, there is no arbitrariness in the choice of the numbers 3 or $\frac{1}{3}$ to express these relations. There really are ratios between the different degrees of these quantities, and these ratios are quite correctly characterized by the rational numbers, no matter how we choose to assign our measuring labels.

So let us start again from the beginning.

3. Numerical Comparison

I begin by introducing a little special terminology. First, when I speak of a *comparative relation* I have in mind primarily any relation which is standardly expressed by means of the grammatical comparative form, i.e. in English by an expression of the grammatical form '. . . is ϕ-er than . . .' or '. . . is more ϕ than . . .'. Associated with every comparative relation there is what I shall call an *equality relation*, and here I have in mind primarily a relation which is expressed in English by the grammatical form '. . . is as ϕ as . . .', while I say that a comparative relation and an equality relation are *associated* if they are, or could be, expressed without ambiguity by expressions grammatically related as '. . . is more ϕ than . . .' (or '. . . is ϕ-er than . . .') and '. . . is as ϕ as . . .'. As schematic expressions for comparative and equality relations that are so associated I accordingly adopt the formulae '. . . is more ϕ than . . .' and '. . . is as ϕ as . . .', which I also abbreviate to '. . . $\succ_\varphi$ . . .' and '. . . $\approx_\varphi$ . . .' respectively. (For technical reasons it proves convenient to have at hand also the relation '. . . $\succcurlyeq_\varphi$ . . .', understood in the obvious way.) Finally, I shall say of any object in the field of a comparative relation '. . . is more ϕ than . . .' that it has (some degree of) ϕ-ness, or more briefly that it is ϕ.

This characterization of comparative and equality relations is so far rather vague and excessively linguistic. We may improve the situation a bit by adding a logical requirement, to ensure that all comparative relations and associated equality relations have the expected ordering properties. To this end let us define

$$\phi a \quad \text{for} \quad (\exists x)(a \succcurlyeq_\varphi x \vee x \succcurlyeq_\varphi a)$$

and stipulate two axioms for comparative relations to ensure that they are all weak ordering relations:

C1 $a \geqslant_\varphi b \ \& \ b \geqslant_\varphi c \rightarrow a \geqslant_\varphi c$

C2 $\varphi a \ \& \ \varphi b \rightarrow a \geqslant_\varphi b \lor b \geqslant_\varphi a.$

The associated equality relation will then be definable by

$$a \approx_\varphi b \quad \text{for} \quad a \geqslant_\varphi b \ \& \ b \geqslant_\varphi a$$

and the comparative relation proper by[16]

$$a \succ_\varphi b \quad \text{for} \quad a \geqslant_\varphi b \ \& \ b \not\geqslant_\varphi a$$

$\leqslant_\varphi$ and $\prec_\varphi$ will be regarded as definitional converses of $\geqslant_\varphi$ and $\succ_\varphi$.

It is easy to check that with these definitions the axioms C1 and C2 provide for all the expected ordering properties, but obviously there are many relations which have the same ordering properties but would not ordinarily be recognized as comparative relations, e.g. the relation '. . . is a man but . . . is not'. It might be possible to formulate further criteria of a logical nature which would rule out a good number of the unwanted cases. For instance it seems to hold of most ordinary comparative relations, but not of the contrived examples, that there is no finite number n such that it can be determined *a priori* that there are just n equivalence classes generated by $. . . \approx_\varphi . . . .$ However, I prefer not to adopt any such further restrictions at this stage, for fear of ruling out some quite acceptable cases (e.g. '. . . is a natural number greater than the natural number . . . but less than 100'). Since our aim is to select those favoured quantities where we can meaningfully speak of ratios, it evidently will not matter if we begin with a very hospitable range of 'quantities'.

Let us come, then, to the question of ratios. Clearly if '. . . is more φ than . . .' is a comparative relation, it will always be grammatically in order to speak of one thing being *twice as φ as* another, or *half as φ*, or *two and a half times as φ*; any expression for a rational number can be coupled in this way with a comparative relation, and the rules of grammar will not be infringed.

---

[16] The point of the clause $a \geqslant_\varphi b$ in the definition of $a \succ_\varphi b$ is to ensure a restriction to the field of the relation $\geqslant_\varphi$. I follow this practice consistently.

However, though such expressions as '. . . is two and a half times as eloquent as . . .' are grammatically in order, we very often want to say that they make no sense, and our problem is to pick out, if possible, the ones that do make sense and to explain how they make sense. Where such an expression does make sense I shall say that it expresses a relation of *numerically definite* comparison (or, for short, of numerical comparison). Strictly speaking, these relations are logically independent of measurement, for (i) as we have seen our ability to measure a quantity does not carry with it an ability to make numerically definite comparisons concerning that quantity and (ii) numerically definite comparisons evidently do not presuppose a choice of unit and therefore do not imply any particular scale of measurement. But the second point is evidently a very trivial one, and discussions of measurement are still relevant to our topic.

In fact we may begin by noticing that writers on measurment commonly single out a special type of measurement, called *fundamental* measurement[17], which is only applicable to certain quantities and which is held to be specially important. It seems to be generally believed that this type of measurement is fundamental *for the scientist*, though so far as I know it has never been satisfactorily explained why it should be of any importance to him, and my discussion of the previous section should at least throw some doubt on the view. I return to the question later (p. 111). However, it *is* very reasonable to hope that the procedure for fundamental measurement will shed light on the non-arbitrary application of numbers to quantities, since the whole point of this procedure is that it relies upon an operation on physical objects which is specifically intended to mirror the arithmetical operation of addition.

An orthodox treatment of fundamental measurement might proceed somewhat like this. We should begin by noting the important properties of the arithmetical operation of addition, which we might perhaps take to be the properties that, where n, m, and p are numbers,

(1) $n + m$ exists and is a number

(2) $n + m = m + n$

[17] I believe that the term originates from Campbell (1920).

(3) $n + (m + p) = (n + m) + p$ *Associative*

(4) $n = m \rightarrow (n + p = m + p)$

(5) $n > m \leftrightarrow (\exists p)(n = m + p)$

Next we say that if a quantity ϕ-ness is to be fundamentally measurable then there must be some combining operation, say $\oplus$, which we may perform upon objects a and b which are ϕ to form a new object $a \oplus b$, called the combination of a and b, which is also ϕ. This operation is to have properties analogous to the properties of addition just noticed, i.e. where a, b, and c are all ϕ we shall require[18]

(1) $a \oplus b$ exists and is ϕ

(2) $a \oplus b \approx_{\varphi} b \oplus a$

(3) $a \oplus (b \oplus c) \approx_{\varphi} (a \oplus b) \oplus c$

(4) $a \approx_{\varphi} b \rightarrow (a \oplus c \approx_{\varphi} b \oplus c)$

(5) $a \succ_{\varphi} b \leftrightarrow (\exists x)(a \approx_{\varphi} b \oplus x)$

If there does exist an operation of combination with these properties it is easy to see how we might use it to explain the sense of numerical comparisons: we simply say that a counts as n times as ϕ as b if and only if a is as ϕ as a combination of n objects which are each as ϕ as b, and further that a is n/m times as ϕ as b if and only if a combination of m objects each as ϕ as a is n times as ϕ as b. Unfortunately, however, we cannot say that this way of introducing a relation of numerical comparison will always yield the right results.

The trouble is that the conditions we have imposed on our combining operation $\oplus$ are not sufficiently exacting to rule out all arbitrariness in the choice of this operation[19]. For instance,

---

[18] It would be usual to include Archimedes' postulate, but I am postponing that to later discussion (§5). With this exception axioms (1)–(5), together with axioms C1 and C2 for comparative relations, are easily seen to be equivalent to those given by Suppes (1951). Ellis (1966, Chap. 5) omits the axiom

$$a \succ_{\varphi} b \rightarrow (\exists x)(a \approx_{\varphi} b \oplus x)$$

and in its place he has a complicated existential condition (stated informally) which in effect requires that if $a \succ_{\varphi} b$ then for any object c, however small, there exists an object x such that the difference in ϕ-ness between a and $b \oplus x$ is less than c.

[19] In another way, however, the conditions are too exacting, as I show on pp. 105, 107.

if we once more consider the measurement of length as an example, the usual operation that would be invoked here is the operation of placing two straight rods end to end in a straight line and taking the *combination* of these rods to be the whole composite rod so formed. However, for all that we have said so far, it appears to be an entirely arbitrary choice that the two rods are to be placed *in a straight line*. For example, if we were to choose instead the operation of placing our rods end to end *at right angles* and taking as their *combination* a rod that just fitted between their free ends (so forming the hypotenuse of the triangle), we should still satisfy the conditions that have been laid down for the combining operation just as well as with the usual choice of this operation[20]. Yet it is clear that the resulting scale for the measurement of length would look quite different from the normal one—in fact n degrees on the old scale would correspond to n^2 degrees on the new scale—and consequently we would find quite different numerical comparisons resulting. Should we, then, follow Ellis in concluding that our present understanding of 'twice as long' turns out to be no less arbitrary than would be any suggestion for understanding 'twice as hot'[21]

I do not see why we should yet be forced to this position, for all that we have shown so far is that the conditions that we have laid down for the combining operation are not sufficient to pick out just one operation, and for this the remedy would seem to be to add further conditions. Furthermore, there is one very obvious way of distinguishing the combining operation that we actually employ from others that might be suggested, and that is that what we in fact understand by the 'combination' of two rods is genuinely their *combination*, i.e. that it is a composite rod and that is actually made up from the two original rods as *parts*, and from nothing else. Also, it is pertinent to remark at this point that the notion of a *part* is really indispensible to a proper statement of an earlier stage of this discussion. For we

[20] The example is taken from Ellis (1966, p. 80) who discusses it very thoroughly. But his conclusions are quite different from mine.

[21] Ellis (*loc. cit.*) thinks that there is an arbitrariness not only in our choice of operation but also in our decision to interpret this operation as reflecting *addition* rather than any other arithmetical operation. However, this seems to be putting the matter back to front, for it seems clear that the only reason we could have for putting any conditions on our operation at all is that we have already decided what arithmetical operation it is to represent.

laid down as our first condition on the operation of combining
that $a \oplus b$ should always exist, and yet a moment's reflection
will reveal that this condition is not satisfied even by the normal
operation of placing rods end to end in a straight line: if a and
b are identical, or if one is a part of the other, or even if they
have any part in common, they evidently cannot be placed end
to end in the way we envisage. (Nor can we stipulate that
$a \oplus a$, for instance, is to be identical with a, for then condition
(v) will not be satisfied). We should therefore have stated that
our five conditions are to hold only where none of a, b, c have
any *part* in common, and so it appears that the notion of a part
is one that we can hardly avoid. But why exactly is it just this
notion that proves so important?

To answer this question we shall, I think, have to approach
our problem from a rather different direction. We are asking for
an interpretation of such numerical comparisons as 'a is n
times as ϕ as b', and our difficulty arises because, for such values
of ϕ as we have considered, any suggested interpretation of this
comparison seems essentially arbitrary. However, there is one
value for ϕ in the context 'twice as ϕ' which leaves us in no
doubt whatever over the interpretation of the phrase, and that
is the case of 'twice as *many*' which is a relation of numerical
comparison based on the comparative relation '. . . more (numerous) than . . .'. For clearly the A's are twice as many as the
B's if and only if the A's may be split into two equal groups
each of which is as many as the B's. For phrases of the form 'n
times as many', then, we may very naturally interpret the
numerical comparison in terms of the relation of subclass to
class, which I would say actually is a case of the relation of part
to whole (provided the null class is excluded from consideration
—see p. 125). Even if it is claimed that the two relations should
be distinguished, it is nevertheless quite clear that they are
extremely similar and differ hardly at all in their formal properties, so that it is at the least a very easy metaphor to say that
the A's are n times as many as the B's if and only if the A's may
be split into n equal parts, each of which is as many as the B's.

However, this criterion, which gives the sense of 'n times' in
the context 'n times as many', also applies unchanged in several
other contexts of the form 'n times as ϕ'. First, the words
'many' and 'much', which are evidently very close in many

ways, are also alike in this feature. I do not see how anyone could object to the suggestion that to say that there is twice as much water in this glass as there is in that glass is the same as to say that the water in this glass may be divided into two equal parts, each of which is as much as the water in that glass. Of course we soon find, when we are comparing amounts of a non-homogeneous stuff or when we are comparing different stuffs, that the relation expressed by 'as much as' has become ambiguous; we must specify whether we mean as much *by weight*, or *by volume*, or perhaps in some other way. Thus the simple notion of being as much as often has to be replaced by the more precise notions of being as heavy as or being as large as, but the same analogy is retained. To say that one object is twice as large as another is still interpreted as being to say that the one contains two discrete parts each of which is as large as the other, and similarly with weight.

At first glance one might suppose that the same account could be applied unchanged to '*n* times as *long*', but here we meet a slight difficulty. It is not quite correct to say that one object is *n* times as long as another if and only if the first can be split into *n* parts each of which is as long as the second; it all depends on how the splitting up is done. Essentially we have to recognize that length is a directional notion, so that what we mean by the length of an object is always its length *in a certain direction* (usually in the direction in which its length is greatest). Consequently the parts that we are to take into account here are, as one might say, parts taken in the same direction, i.e. parts obtained by dividing the object in a plane perpendicular to that direction. Finally, the length of these parts that is in question when they are compared with the second object is again to be understood as their length in that same direction. But this complication, though tedious to state, is surely very easy to understand, and does not disturb the general lines of our approach. All that is required is that the notion of a part should be understood in a slightly special way in this case, so that only certain of the parts of the object are allowed to count. But still it does seem to be the notion of a part that is important for understanding those 'combining' operations that underlie our actual numerical comparisons.

Actually we can at this point drop the notion of a combining operation altogether, which is of some advantage, as we have still not exhausted the difficulties it involves. I have already pointed out that if our first condition, requiring that $a \oplus b$ always exists, is to be satisfied in the sense we intend, then it must be limited to the case when a and b have no part in common. But there is still a further difficulty with this condition. The idea was that we should be able to combine *any* two (discrete) objects that were ϕ, and in the case of length we decided that the combining operation was to consist of putting the two objects end to end in a straight line. However in the subsequent discussion I concealed a difficulty by speaking only of applying the operation to *rods*, and it was easy enough to accept that *any* two (discrete) rods can be placed end to end. But obviously it is not true that any two objects whatsoever that have length can be placed end to end (e.g. two roads) and so we never did in fact satisfy this condition. However, now that we have seen that it is the notion of a part that is important, the difficulty can easily be overcome. Instead of considering the operation of putting together out of parts it is altogether more straightforward to concentrate instead on the parts themselves, for what matters is just what parts a thing has and not what can be done to it or to them. Accordingly it seems better to put our suggested explanation of numerical comparisons in the form[22]

a is n times as ϕ as b if and only if a contains and is exhausted by n mutually discrete parts, each of which is as ϕ as b.

The sole rider so far is that the notion of a part may have to be understood rather specially in some cases, e.g. length. (I shall characterize this special understanding more fully in the next section.) By analogy with the notion of fundamental measurement I propose to call numerical comparisons that can be explained in this way *fundamental* numerical comparisons, because they best preserve the all-important analogy to the central case of 'n times as many as'. However, not all numerical comparisons are fundamental.

There are some cases of numerical comparison which as they stand do not fit the criterion just stated, but which can easily be brought into harmony with it by a little rewriting. For

[22] The suggestion, of course, applies only where n is a natural number. Further elaboration will be needed for other cases.

example, if it is said that a is *twice as far* as b (from here), this obviously has nothing to do with what parts a has, but we can easily rewrite the statement so that the criterion does apply. What is meant is that *the distance* from a to here is twice as great (twice as far) as the distance from b to here, and distances may well be thought of as things which have parts in the relevant way. (We may wish to add that distances and their parts should somehow be viewed as logical constructions from the physical objects that could be laid over those distances and *their* parts, but that does not concern us here.) In this case the obviousness of the transformation no doubt helps to explain why we do not consider that there is any arbitrariness in such a use of 'twice'.

A superficially similar case is that of '*twice as fast*'. It is very natural to explain that a is travelling twice as fast as b if and only if a is travelling *twice as far* as b in equal times (and we may put the criterion counterfactually if we wish to include instantaneous velocities). However, the fact that speeds are explained in terms of distances *and times* in fact makes this case rather different. Admittedly the connections between the concepts of speed, distance, and time are very close in our ordinary thinking, and it is plausible to regard the criterion just given as analytic for the ordinary man. In other formally similar cases, however, this is not so. For instance if it is explained that one mechanical hoist is *twice as powerful* as another because it can lift an equal weight *twice as fast*, I think that this is apt to strike most of us as a somewhat artificial stipulation and certainly not to be extracted from an analysis of the ordinary concept of power. In each case, however, numerical comparisons in respect of one quantity are being explained by correlating that quantity with some others, for as we say (somewhat schematically):

$$\text{speed} = \frac{\text{distance}}{\text{time}}$$

$$\text{power} = \frac{\text{resistance overcome} \times \text{distance}}{\text{time}}.$$

Is there any reason why most people think the first correlation entirely natural and the second highly artificial?

One point that could be made is that the comparative relation 'faster than' can presumably be understood in advance of any such correlation, since *a* is travelling faster than *b* if *a* overtakes *b*, whereas it is difficult to think of an independent way of explaining 'more powerful than'. In this latter case it seems that the correlation is needed even to settle the *order* generated by the comparative relation, which seems to show that the ordinary pre-theoretical understanding of power is crucially vague. So let us turn to another example.

An interesting case is that of brightness, or intensity of illumination. We can judge perfectly well by eye whether one light is brighter than another, or whether one surface is more brightly illuminated than another, so the order generated by the relation 'brighter than' can be determined in advance of any correlations between brightness and other quantities. However, the eye seems to tell us nothing about being twice as bright. To remedy this lack we may perhaps be told that when two lights *a* and *b* are so placed that they illuminate the same surface equally, then if *a* is *n* times as far from the surface as *b* is, *a* is to be counted as n^2 times as bright as *b* is. Generally,

$$\frac{\text{brightness of light}}{\text{intensity of illumination}} = (\text{distance})^2.$$

In default of theoretical considerations concerning the nature of light and its emission and dispersal (and perhaps even with those theoretical considerations), this stipulation is apt to strike one as entirely arbitrary. Why should we not say more simply that the brightness varies directly as the distance, instead of using the square of the distance? If we were to adopt this course, 'twice as bright' would be understood in a way very similar to 'twice as fast', and it seems quite possible that we should in the end come to accept both as equally 'natural'. But, at least at first glance, there seems nothing very natural about the square of the distance.

However our actual understanding of 'twice as bright' has quite a different basis, which makes it very similar indeed to a fundamental numerical comparison. We do not ordinarily think of the brightness of a light as being the sum of the brightnesses of its parts, since we do not think of a light as something which must have parts that are lights, and actually separating a light into its parts is apt to leave no lights at all. But in certain cases we *can* view the situation in this way. If we take two lights of equal brightness (for example two standard candles) and place them close to one another, we are prepared to view the composite object so formed as being itself a light, and clearly it will be a light that is brighter than either of its component lights. So if in this case we call the composite light twice as bright as each of its components we are preserving the analogy with our paradigm cases sufficiently for us to feel that this is not just an arbitrary use of 'twice'. But then we shall at once extend the usage of 'twice as bright' beyond the case where we are actually dealing with composite lights by relying on the principle that if *a* is twice as bright as *b*, and *c* is as bright as *a*, then *c* also must be twice as bright as *b*. And thus it comes about that in the end we understand '*a* is twice as bright as *b*', to mean something like '*a* is as bright as any composite light that has as components two lights which are each as bright as *b*', where it is not implied that *a* itself is a composite light having parts which are lights. Obviously the analogy with our central cases has not been lost to view.

On reflection it does not seem possible to lay down any general principles to govern our extension of numerical comparisons beyond the fundamental cases, and I think that there is nothing to be gained from attempting a ruling on when a type of numerical comparison counts as arbitrary and when it counts as natural. There are different ways of extending the central analogy, e.g. *any* way of satisfying the standard axioms for fundamental measurement will preserve *some* features of that analogy, and no doubt some will strike us as more natural extensions than others. Also there are different cases where some numerical comparisons are used to define others, some of which seem now incorporated into our understanding of the concepts involved (e.g. speed, and perhaps density), while others remain more or less artificial technical uses (e.g. power, and perhaps loudness). However, I do claim that what I call the *fundamental* numerical comparisons deserve to be called natural, because they preserve a very close analogy with the undisputed case, and further that in these fundamental cases we have the basis for the whole enterprise of applying numbers to quantities. Accordingly I concern myself from now on only with the fundamental cases.

As I remarked earlier, it should not be supposed that these fundamental numerical comparisons are in any way fundamental *for science*. As it happens, the scales of measurement in scientific use are very often derived in one way or another from fundamental numerical comparisons, but there is no *a priori* reason why this should be so, and there is no reason (except conservatism) why it should remain so. In particular, it would be a mistake to suppose that scales based on fundamental numerical comparisons are in any way more 'informative' than other scales[23], for any mathematically definite transformation from one scale to another must preserve the same 'information'. (What happens is just that the same physical relation or operation is represented now by one and now by another arithmetical relation or operation.) Nor is there any reason to think that the standard accounts of fundamental *measurement* (which merely require *some* operation with properties analogous to addition) single out any important feature of a quantity. For example, it is possible to devise a procedure for measuring temperature which satisfies the criteria for fundamental measurement[24], but no interest—scientific or otherwise—attaches to this. Hence I think that what is fundamental about so-called fundamental measurement has been misunderstood. What is *aimed at* by the standard accounts is really what I call fundamental numerical comparison, but they do not reach this target since the crucial notion of a *part* is altogether omitted. No doubt the reason why it is omitted is that there is no reason to suppose that it is of any noticeable scientific importance, except perhaps in the somewhat parochial case of geometrical quantities. But scientists need to measure a very wide variety of quantities.

We could have arrived at the notion of a part from a different

---

[23] *Pace* Ellis 1966, p. 100. Contrast p. 62.

[24] The example is due to Ellis (1966, pp. 101–3). Strictly it is temperature intervals that are measured, and our 'objects' are pairs of bodies connected by a thermocouple, say $\langle x, y \rangle$ with the hotter body x always written first. We have $\langle x, y \rangle \approx \langle z, w \rangle$ if and only if no current flows in the coupling $x - y - w - z$, and we have $\langle x, y \rangle \oplus \langle z, w \rangle \approx \langle a, b \rangle$ if and only if no current flows in the coupling $x - y - z - w - b - a$. The resulting scale is by no means the usual one.

direction, by considering rational numbers not as *ratios* but as *fractions*. It is obvious that in contexts such as 'half an orange' or 'two-thirds of a sheep' we are concerned with *parts*, and a little reflection would reveal that a comparative relation is also involved. (It may make a difference whether we want two-thirds by *weight* or by *volume*.) However, the weakness with this approach is that, although the point emerges clearly enough for *proper* fractions, other 'vulgar' fractions are not naturally used in this role, so our conclusion might have seemed more limited than in fact it is. However, as matters stand we may justifiably add this point as further support to the conclusion that we have reached by a different route: to understand the application of rational numbers we must focus on the idea of a part. Accordingly in the next section I consider parts independently of their connection with numbers, and then I return to a more detailed discussion of fundamental numerical comparison.

4. Parts

The formal properties of the relation of part to whole have been studied and axiomatized in the system known as 'mereology' or the 'calculus of individuals'[25]. The axioms I shall present here are, however, somewhat weaker than the usual ones. I shall first present my own system—which I call the system P (for Parts)—and then compare it with the standard system of mereology and with the closely similar system of Boolean algebra, giving my reasons for preferring the weaker axioms.

The most obvious formal properties of the part–whole relation are that it is reflexive, antisymmetric, and transitive. Accordingly, where '$\ldots \subset \ldots$' represents this relation, we shall certainly expect to find the following theses forthcoming:

(1) $a \subset a$

(2) $a \subset b \;\&\; b \subset a \rightarrow a = b$

(3) $a \subset b \;\&\; b \subset c \rightarrow a \subset c$.

However, it is obvious that more than these properties will be needed. We may make some progress here by introducing the

---

[25] See Goodman 1951, Chap. II, § 4; Goodman and Leonard 1940; Woodger 1937, Appendix E (by Tarski).

notion of discreteness, for two objects are said to be discrete if and only if they have no part in common. If we abbreviate '*a* is discrete from *b*' by *a*|*b* we shall therefore define

D1 $a|b$ for $\sim(\exists x)(x \subset a \,\&\, x \subset b)$.

With the help of this notion we can now formulate a further property of the part–whole relation. For if first *a* and *b* have some part in common, and this part is a proper part of *a*, then we may 'subtract' *b* from *a* and still leave something of *a* remaining; and if second *a* and *b* have no part in common, then trivially we may 'subtract' *b* from *a* and still leave the whole of *a* remaining; and in each case the remainder will be a part of *a* and discrete from *b*. There will be no such remainder only if what *a* and *b* have in common is the whole of *a*, and in that case *a* will itself be a part of *b*. So we may assert

(4) $a \not\subset b \rightarrow (\exists x)(x \subset a \,\&\, x|b)$.

The four theses so far listed constitute the basic properties of the part–whole relation. We may conveniently take the second of them as a *definition* of the identity appropriate to this system, defining

D2 $a = b$ for $a \subset b \,\&\, b \subset a$

for it is clear that if *a* and *b* are identical according to this definition they will be intersubstitutible in all contexts built up from the part–whole relation. (Whether they will also be intersubstitutible in other contexts does not yet concern us.) The remaining three theses are easily seen to be equivalent to the single axiom

P1 $a \subset b \leftrightarrow (\forall x)(b|x \rightarrow a|x)$

which I therefore adopt.

In addition to this 'universal' axiom we shall also require axioms specifically providing for existence. These axioms are used in conjunction with the notions of the *sum, product,* and *difference* of several objects. The sum, $a+b$, of *a* and *b* is of course that whole which has just *a* and *b* as parts and nothing else discrete from them. In fact anything discrete from both *a* and *b* must also be discrete from their sum, so we have

$$a \subset a+b \,\&\, b \subset a+b \,\&\, (\forall x)(a|x \,\&\, b|x \rightarrow a+b|x)$$

i.e., in view of P1,

$$'(\forall x)(a + b|x \leftrightarrow a|x \ \& \ b|x).$$

Another equivalence that follows from this one is

$$(\forall x)(a + b \subset x \leftrightarrow a \subset x \ \& \ b \subset x)$$

and either may be used to define the sum. For definiteness we may choose the second, and to construct a formal definition we may employ a Russellian definite description:

D3 $a + b$ for $(\imath y)(\forall x)(y \subset x \leftrightarrow a \subset x \ \& \ b \subset x).$

(The identity concealed in this definite description is of course to be taken as the identity defined by D2.) In the same way the *product*, $a \cdot b$, which is the greatest common part of a and b, is defined dually:

D4 $a \cdot b$ for $(\imath y)(\forall x)(x \subset y \leftrightarrow x \subset a \ \& \ x \subset b).$

Finally the *difference* $a - b$, which is as it were what remains of a when b is 'subtracted', is defined:

D5 $a - b$ for $(\imath y)(\forall x)(x \subset y \leftrightarrow x \subset a \ \& \ x|b).$

Since sum, product, and difference are introduced as definite descriptions, it is necessary to say something of the *scope* of these descriptions. And since it would be excessively tedious to cumber the notation with explicit scope indications, it is convenient to adopt some convention for omitting them. For this purpose I stipulate that the scope of a definite description should always be understood to be the smallest possible. To make this ruling definite, let us regard the binary relation $\ldots \subset \ldots$ as the sole primitive predicate, so that all occurrences of $|$ or $=$ are to be paraphrased in accordance with D1 and D2 before the description is eliminated. Then the ruling is that where ζ is any *term* (i.e. any variable or definite description)

$$(\imath y)(-y-) \subset \zeta$$

abbreviates

$$(\exists x)((\forall y)((-y-) \leftrightarrow y = x) \ \& \ x \subset \zeta)$$

and where ξ is any *variable*

$$\xi \subset (\imath y)(-y-)$$

abbreviates

$$(\exists x)((\forall y)((-y-) \leftrightarrow y = x) \,\&\, \xi \subset x).$$

Thus for formal purposes the definite description counts as defined only where its scope is minimal. In later developments we shall also wish to use our definite descriptions in conjunction with comparative relations, and for this purpose the above abbreviations are to be regarded as applying also when a comparative relation $\succcurlyeq_\varphi$ is substituted for $\subset$, but all other uses of these descriptions are officially forgone. The point of this manoeuvre is simply that our formulae would otherwise become ambiguous, since we shall not assume that our sums, products, and differences always exist.

Let us return, then, to the question of existence. It is easily seen from D2 that at most one object can satisfy the condition for being the sum, product, or difference of two others, but we still need to be assured that there will in appropriate circumstances be at least one such object. Accordingly I propose to adopt the following existential postulates:

(5) $(\exists x)(a \subset x \,\&\, b \subset x) \rightarrow (\exists y)(\forall x)(y \subset x \leftrightarrow a \subset x \,\&\, b \subset x)$

(6) $(\exists x)(x \subset a \,\&\, x \subset b) \rightarrow (\exists y)(\forall x)(x \subset y \leftrightarrow x \subset a \,\&\, x \subset b)$

(7) $(\exists x)(x \subset a \,\&\, x|b) \rightarrow (\exists y)(\forall x)(x \subset y \leftrightarrow x \subset a \,\&\, x|b).$

In view of the uniqueness just noted, together with the obvious fact that these implications are reversible, the postulates could equally be expressed as

(5') $(\exists x)(a \subset x \,\&\, b \subset x) \leftrightarrow \mathrm{E}!(a + b)$

(6') $(\exists x)(x \subset a \,\&\, x \subset b) \leftrightarrow \mathrm{E}!(a \cdot b)$

(7') $(\exists x)(x \subset a \,\&\, x|b) \leftrightarrow \mathrm{E}!(a - b).$

(Here E! is standard Russellian notation, and is defined

$$\mathrm{E}!(\imath y)(-y-) \quad \text{for} \quad (\exists x)(\forall y)((-y-) \leftrightarrow y = x)).$$

The existential postulate concerned with differences is obvious enough. For the difference $a - b$ (which we may read as 'a without b') is precisely the part of a which remains when b is subtracted, and that there is such a part under the conditions proposed was already implicit in the informal argument given for (4) above. As for the postulate concerned with products, this simply states that if a and b have a common part at all then they have a greatest common part, while the postulate for sums states analogously that if a and b have a 'common whole' then they have a 'least common whole'. This last postulate is much weaker than the corresponding postulate in orthodox mereology, for mereology here makes the simplifying assumption that the sum $a + b$ always exists. My postulate may well seem altogether too weak, and I discuss the point further below (pp. 122–8).

The system introduced so far, with existential postulates (5)–(7), may be called the *elementary* system P. It differs little from an elementary system of Boolean algebra, except of course that its existential postulates are much less demanding. Consequently within this system one can prove most of the standard theorems of Boolean algebra provided that they are prefaced by sufficient existential conditions to assure the existence of the terms involved. For example, whereas in Boolean algebra we have the unconditional principle of distribution

$$a \cdot (b + c) = (a \cdot b) + (a \cdot c)$$

in the system P we have to content ourselves with

$$\mathrm{E}!(a \cdot b) \ \& \ \mathrm{E}!(a \cdot c) \ \& \ \mathrm{E}!(b + c) \rightarrow a \cdot (b + c) = (a \cdot b) + (a \cdot c)$$

Apart from this complication the usual proofs can often be transferred without difficulty, and where they cannot this generally arises because in ordinary Boolean algebra (and in ordinary mereology) it is assumed that there exists a universal element 1, so that each term a has a complement $1 - a$ (usually written a'), which is very convenient. The system P assumes no such universal element, however, and so makes no use of complements, which again leads to complexities in the proofs. But usually what can be accomplished in Boolean algebra by means of complements can also be accomplished in the system P by means of differences, so there is still no great divergence in the theorems. However we have not yet finished with our presentation of the system, for the *elementary* system that we have considered so far is not yet adequate. I now proceed, therefore, to the *extended* system P, which corresponds rather to what is called a *complete* Boolean algebra.

The notions of sum and product that we have so far defined allow us to speak of the sum and product of any two given objects, and therefore (by repeated applications) of the sum and product of the objects given in any (finite) list. However, it is extremely useful—indeed for further developments essential—to be able to speak of the sum and product of all objects satisfying a given condition, whether or not we are able to list those objects. For this purpose the definitions of sum and product need to be rephrased in a more inclusive way, and the most natural procedure would seem to be to define 'the sum of all the x's such that Fx' and 'the product of all the x's such that Fx' respectively[26]

D6 $\Sigma\{x:Fx\}$ for $(\imath y)(\forall z)(y \subset z \leftrightarrow (\forall x)(Fx \rightarrow x \subset z))$

D7 $\Pi\{x:Fx\}$ for $(\imath y)(\forall z)(z \subset y \leftrightarrow (\forall x)(Fx \rightarrow z \subset x))$.

It is clear that these definitions preserve our original understanding of sums and products, in so far as the product of all the F's is defined as the greatest common part of all the F's, while the sum of all the F's is defined dually as the least common whole of all the F's. Also, it is easily seen that these definitions contain our original definitions in the sense that substitution of $x = a \lor x = b$ for Fx at once yields the previous definitions of $a + b$ and $a \cdot b$.

Now we have to consider the conditions under which these newly defined sums and products exist. In the case of products it is clear that, as we stipulated earlier that a and b will have a greatest common part provided that they have a common part at all, so now we should say that all the F's will have a greatest common part provided that they do have some common part. However it should be noticed that the most natural way of formulating this condition, namely

$$(\exists y)(\forall x)(Fx \rightarrow y \subset x)$$

permits the condition to be *satisfied* in the case where there are no F's at all, but it seems to me reasonable to say that the

[26] I have found it convenient here (and elsewhere) to use the class-notation $\{x:Fx\}$, but it should be noted that throughout this book the use of this notation is syncategorematic; I make no assumption whatever about the existence of classes.

product will not exist in this case, because if there are no F's at all then surely the F's do not have a common part in any ordinary sense. Consequently the existential postulate needed for products in general should be formulated as

(8) $(\exists x)(Fx)$ & $(\exists y)(\forall x)(Fx \to y \subset x) \to$

$(\exists y)(\forall z)(z \subset y \leftrightarrow (\forall x)(Fx \to z \subset x))$

and it turns out that this is interdeducible with the analogous existential postulate for sums, namely

(9) $(\exists x)(Fx)$ & $(\exists y)(\forall x)(Fx \to x \subset y) \to$

$(\exists y)(\forall z)(y \subset z \leftrightarrow (\forall x)(Fx \to x \subset z))$.

(The interdeducibility arises because a sum may clearly be taken to be a product of common wholes, and a product may clearly be taken to be a sum of common parts.)

In the presence of the extended postulates (8) or (9) the elementary postulates (5) and (6) are of course superfluous. But the elementary postulate (7) dealing with differences is not superfluous, and at the moment is still required axiomatically. It is worth noting, though, that if we make one modification in our extended existential postulate then (7) will become deducible. I mentioned that the elementary sum $a + b$ could have been defined in terms of discreteness, but on the present definition we need postulate (7) in order to show that

$$\mathrm{E}!(a + b) \to (\forall x)(a + b | x \leftrightarrow a | x \ \& \ b | x).$$

In a similar way we may use postulate (7) to show from D6 that

$$\mathrm{E}!\Sigma \{z : Fz\} \to (\forall x)(\Sigma \{z : Fz\} | x \leftrightarrow (\forall z)(Fz \to z | x)).$$

Consequently we could equally well have proceeded instead by defining the extended sum in terms of discreteness. In that case the natural existential postulate to provide for the existence of sums would have been

P2 $(\exists x)(Fx)$ & $(\exists y)(\forall x)(Fx \to x \subset y) \to$

$(\exists y)(\forall z)(y | z \leftrightarrow (\forall x)(Fx \to x | z))$.

Now it turns out that in the presence of P1 this postulate P2 is interdeducible with (7) and (8), or (7) and (9), and so will suffice as our sole existential postulate. I therefore adopt it.

The axiomatic basis of the system P is now complete. To recapitulate briefly, there is first a definition

D1 $a|b$ for $\sim(\exists x)(x \subset a \,\&\, x \subset b)$

and a universal axiom

P1 $a \subset b \leftrightarrow (\forall x)(b|x \rightarrow a|x)$.

This universal axiom is equivalent to the conjunction of

P1(a) $\begin{cases} a \subset a \\ a \subset b \,\&\, b \subset c \rightarrow a \subset c \end{cases}$

and

P1(b) $a \nsubseteq b \rightarrow (\exists x)(x \subset a \,\&\, x|b)$.

The system is completed with an existential axiom

P2 $(\exists x)(Fx) \,\&\, (\exists y)(\forall x)(Fx \rightarrow x \subset y) \rightarrow$

$(\exists y)(\forall z)(y|z \leftrightarrow (\forall x)(Fx \rightarrow x|z))$

which (in the presence of P1) is equivalent to the conjunction of

P2(a) $(\exists x)(Fx) \,\&\, (\exists y)(\forall x)(Fx \rightarrow x \subset y) \rightarrow$

$(\exists y)(\forall z)(y \subset z \leftrightarrow (\forall x)(Fx \rightarrow x \subset z))$

and

P2(b) $(\exists x)(x \subset a \,\&\, x|b) \rightarrow (\exists y)(\forall x)(x \subset y \leftrightarrow x \subset a \,\&\, x|b)$.

Clearly P1(b) and P2(b) could be amalgamated to form the single axiom

P(b) $a \nsubseteq b \rightarrow (\exists y)(\forall x)(x \subset y \leftrightarrow x \subset a \,\&\, x|b)$

so we could, if preferred, adopt P1(a), P2(a), and P(b) as our axioms. Let us now turn to a comparison with other related systems.

Any relation which satisfies P1(a) generates a partial order (with its identity defined as in D2). A relation which in addition satisfies an *unrestricted* version of P2(a), namely

B2(a) $(\exists y)(\forall z)(y \subset z \leftrightarrow (\forall x)(Fx \rightarrow x \subset z))$

is said to generate a (complete) *lattice*. For such a relation we can evidently define sums (least upper bounds) and products (greatest lower bounds) as in the system P, and these sums and products will *always* exist. In addition there will exist a null element 0 and a universal element 1 such that

$$(\forall x)(0 \subset x), \quad (\forall x)(x \subset 1)$$

(as we see by substituting say $x \neq x$ and $x = x$ for Fx in B2(a)). Consequently in these cases the discreteness relation defined by D1 is never satisfied, but what was previously intended by $a|b$ can now be expressed instead as $a \cdot b = 0$.

Now a (completely) *distributive* lattice will also satisfy an unrestricted version of P2, namely

B2 $(\exists y)(\forall z)(y \cdot z = 0 \leftrightarrow (\forall x)(Fx \to x \cdot z = 0))$.

This may be seen as follows. If the lattice is distributive it will hold that

$$a \cdot (b + c + d + \ldots) = (a \cdot b) + (a \cdot c) + (a \cdot d) + \ldots$$

and more generally that

$$a \cdot \Sigma\{x : Fx\} = \Sigma\{y : (\exists x)(y = a \cdot x \;\&\; Fx)\}$$

or, as we may more briefly say,

$$a \cdot \Sigma\{x : Fx\} = \Sigma\{a \cdot x : Fx\}.$$

However, we know from B2(a) that

$$\Sigma\{a \cdot x : Fx\} \subset 0 \leftrightarrow (\forall x)(Fx \to a \cdot x \subset 0).$$

Putting these last two together we have

$$a \cdot \Sigma\{x : Fx\} \subset 0 \leftrightarrow (\forall x)(Fx \to a \cdot x \subset 0)$$

whence B2 follows at once. Thus a distributive lattice will satisfy axioms corresponding to P1(a) and P2. However it will not necessarily satisfy the axiom corresponding to P1(b), namely

B1(b) $a \nsubseteq b \to (\exists x)(a \cdot x \neq 0 \;\&\; b \cdot x = 0)$.

When this last axiom is also satisfied, so that both B1 and B2 are satisfied, we in fact have a (complete) Boolean algebra[27].

---

[27] See for example Tarski 1956.

The system P, then, differs from a Boolean algebra just in the fact that its existential postulate is rather heavily restricted, while the corresponding postulate for Boolean algebra has no restrictions. The standard system of mereology is intermediate between the two, for it has the same universal axiom P1 and the existential axiom

M2 $(\exists x)(Fx) \rightarrow (\exists y)(\forall z)(y|z \leftrightarrow (\forall x)(Fx \rightarrow x|z))$.

The restriction that the sum of all the F's should exist only when there are some F's is a fairly weak restriction, and it is easy to see that the only real difference it makes is that mereology lacks a null element. More precisely, suppose we begin with a pair of axioms which are non-committal over the null element, say P1 and M2 with D1 given in the neutral form

$a|b$ for $(\forall x)(x \subset a \,\&\, x \subset b \rightarrow (\forall y)(x \subset y))$.

Then if we postulate a null element separately, i.e.

$$(\exists x)(\forall y)(x \subset y)$$

the result is an axiomatization of Boolean algebra, while if we postulate instead that there is no null element, i.e.

$$\sim(\exists x)(\forall y)(x \subset y)$$

the result is an axiomatization of mereology (with the extra but uninteresting information that there is more than one element).

The system P is like mereology in denying the existence of a null element, and the difference between them is that P is non-committal over the existence of a number of other elements which mereology assumes. The most conspicuous example is the universal element, and if we add to P a postulate asserting the existence of this element, i.e.

$$(\exists x)(\forall y)(y \subset x)$$

then the result is once more an axiomatization of mereology. However it should be noted that P does not *deny* the existence of a universal element; it is not incompatible with mereology, but simply weaker, and it makes no assumptions which are not also assumed in mereology.

Clearly what is characteristic of P is its weak existential axiom, and the nature of this axiom may emerge more clearly if we turn to a comparison with set theory. Suppose, then, that the variables of the system P are interpreted as ranging over all non-null sets of a set theory, and that the relation '... $\subset$ ...' is interpreted as inclusion between sets, defined in the usual way:

$$a \subset b \quad \text{for} \quad (\forall x)(x \in a \rightarrow x \in b).$$

Let us also assume, for simplicity, that the set theory already provides for the existence of all unit sets, i.e. that it already contains the theorem

$$(\exists y)(\forall x)(x \in y \leftrightarrow x = a).$$

Then it is easily seen that the set theory must already verify P1, and it also turns out that the effect of adding P2 is precisely the effect of adding an axiom of subsets (axiom of separation, *aussonderung*). For P2, when its variables are interpreted as restricted to non-null sets, may be expressed with unrestricted variables (and taking into account the definition of discreteness) as

$$(\exists x)(\exists w)(w \in x \,\&\, Fx) \,\&\, (\exists y)(\exists w)(w \in y \,\&\, (\forall x)(\forall w)(w \in x \,\&\,$$
$$Fx \rightarrow x \subset y)) \rightarrow (\exists y)(\exists w)(w \in y \,\&\, (\forall z)(\forall w)(w \in z$$
$$\rightarrow (\sim(\exists w)(w \in y \,\&\, w \in z) \leftrightarrow (\forall x)(\forall w)(w \in x$$
$$\&\, Fx \rightarrow \sim(\exists w)(w \in x \,\&\, w \in z))))).$$

Now if in this somewhat unwieldy formula we substitute for Fx the expression $(\exists y)(Fy \,\&\, (\forall z)(z \in x \leftrightarrow z = y))$, then the resulting thesis simplifies by fairly obvious transformations to

$$(\exists x)(Fx) \,\&\, (\exists y)(\forall x)(Fx \rightarrow x \in y) \rightarrow (\exists y)(\forall x)(Fx \leftrightarrow x \in y).$$

And this is equivalent to our original, since the deduction may be reversed by substituting $(\exists y)(x \in y \,\&\, Fy)$ for Fx in this second formula. Now, since the usual axiom of subsets may be written

$$(\exists y)(\forall x)(Fx \rightarrow x \in y) \rightarrow (\exists y)(\forall x)(Fx \leftrightarrow x \in y)$$

it is clear that the axiom P2 is under this interpretation just the axiom of subsets restricted to non-null sets. Of course under the same set-theoretical interpretation the unrestricted existential postulate B2 for Boolean algebra becomes the unrestricted and inconsistent principle of abstraction

$$(\exists y)(\forall x)(Fx \leftrightarrow x \in y)$$

and the slightly restricted existential postulate M2 for mereology becomes the slightly restricted principle of abstraction

$$(\exists x)(Fx) \rightarrow (\exists y)(\forall x)(Fx \leftrightarrow x \in y)$$

which is equally unacceptable. (Though not formally inconsistent, it implies that the universe contains only one object[28].) It follows that Boolean algebra and mereology are systems too strong to be given this set-theoretical interpretation, whereas the interpretation is possible for P because of its weak existential axiom.

So much, then, for the formal liaisons of the system P with other systems. By comparison with mereology and Boolean algebra it is weak, and the existential conditions which have to be prefaced to most of its theorems are undeniably somewhat tedious to work with. The compensating advantage is that it is capable of a much wider range of interpretations. But one might ask whether this really is an advantage. After all, it may be said, we were supposed to be axiomatizing the relation of part to whole, and so what we should be trying to do is to capture precisely those truths that hold of this particular relation. If our axiomatic system happens to fit some other relations too, why should this be thought any sort of advantage? To this I would reply (a) that the objection assumes a premise which is at least not obviously correct, *viz.* that there is just *one* relation which is *the* relation of part to whole, and (b) that anyway my present purpose is in fact better served by allowing several distinct relations to count as part–whole relations.

There is no doubt that we do use the notion of a part in the most diverse range of contexts. Ordinary physical objects are said to have parts which are themselves physical objects;

---

[28] Briefly, the argument is as follows: (i) by substituting $x \notin x$ for Fx we deduce $(\forall x)(x \in x)$; (ii) by substituting $x = a \lor x = b$ for Fx and using (i) we deduce $(\forall xy)(x \in y \lor y \in x)$; (iii) by substituting $x \notin a$ for Fx and using (ii) we deduce $(\forall xy)(x \in y)$; (iv) by substituting $x = a$ for Fx and using (iii) we deduce $(\forall xy)(x = y)$.

somewhat less substantial 'geometrical objects'—lines, triangles, spheres—are said to have equally geometrical parts; and spatial 'objects' more abstract still—lengths, areas, volumes— may also be said to have parts. The case is similar with processes, stretches of time, and durations. And these are all pretty ordinary uses of the word. But we also speak without qualms of the parts of any sort of collection, even when the collection cannot itself be viewed as spatially or temporally divisible; for example, the notes of which a musical chord is made up may be called its parts or (in a somewhat different vein) a body of doctrine may be said to contain this or that particular thesis as a part. These do not seem to be obviously metaphorical uses; at any rate they are very much more literal than, say, 'Part of the deadlock was due to . . .' (which means only 'The deadlock was partly due to . . .'). But the bounds of metaphor are not easy to determine. For example if we list the first person singular present, the infinitive, the first person singular perfect, and the gerund of a Latin verb as its 'principal parts', are we implying that the verb itself is somehow 'made up out of' these and others of its inflexions? And would the implication be correct?

In view of this very wide range of uses, I think it is unprofitable to concentrate upon some one relation which is supposed to be *the* relation of part to whole, and I view the system P rather as an attempt to capture the formal features shared by all relations which are reasonably regarded as part–whole relations. Of course there is an element of stipulation in this, for there is no very objective way of determining whether a relation which fails to satisfy the axioms of P is nevertheless *reasonably* regarded as a part–whole relation. However, my justification for the stipulation is that it is the properties given by P that are required if the part–whole relation is to serve as a basis for numerical comparison in the way sketched in the last section and enlarged upon in the next. But before we come to this it will be well to consider briefly which relations are permitted as part–whole relations by my stipulation. In this connexion the salient features of P that should engage our attention are (i) the insistence that there should *not* exist a null element, and (ii) the insistence that there *should* exist sums, products, and differences in the appropriate circumstances.

The existence of a null element, i.e. an element that is a part of every element, is surely incongruous for the vast majority of part–whole relations. The most interesting exception is the relation of inclusion between classes or sets, and this is a somewhat crucial case since on my account it is the analogy between this relation and more concrete part–whole relations that explains why the notion of a part is at all important to us. However, it is now time to make some finer distinctions, for in fact we do *not* very naturally use the terminology of parts when speaking of sets themselves; we do not very naturally say that one *set* is a part of another *set* at all. What we do say is that the *membership* of the one is a part of the *membership* of the other, or that the *members* of the one form part of the *members* of the other, and this way of speaking does rule the null set out of consideration. For since the null set has no members and no membership, neither its members nor its membership can be a part of anything. The locution also indicates that, properly speaking, the part–whole relation involved here is one that holds between the members or membership of one set and the members or membership of another, and we may perhaps take the terms of this relation to be what Russell called 'classes as many' as opposed to 'classes as one'[29]. But if we have any reason to posit sets (i.e. 'classes as one') in addition to their memberships, then there need be no objection to viewing the derived relation '. . . is not null and is included in . . .' as itself a part–whole relation. We now see that the restriction to non-null sets simply reflects the fact that this *is* a derived relation. Similar remarks apply to the corresponding second-level relation holding directly between first-level predicates, *viz.* the relation $(\forall x)((\ldots x \ldots) \to (-x-))$, which will still be available to us even if we are sceptical about the existence of sets[30]. No doubt, in view of our imperfect understanding of Russellian classes as many, we shall for formal purposes prefer to use one or other of the derived relations.

Let us come now to the question of the existence of sums, products, and differences. Here the weakness of our existential

[29] Russell 1903, especially Chap. VI.

[30] Here I refer to my remarks on pp. 128–9 concerning type-neutrality. Relations between objects (*viz.* sets), between classes as many, and between predicates must presumably be relations of different logical types; but they may all be part–whole relations.

postulate is of some advantage, as we saw when considering its set-theoretical interpretation. A stronger postulate might have caused some difficulty here, and possibly in other areas too, perhaps because it forced the existence of sums that were too large or of sums that were too heterogeneous in composition. However, I should admit that even my weak postulate has one feature that is not perhaps altogether intuitive. If we consider an interpretation of P in which our elements are spatial areas or volumes, then by applying the existential postulate we shall be able to generate new elements which one would not ordinarily call *an* area, or *a* volume, because they are discontinuous, i.e. they are composed of sub-areas or sub-volumes which are not in any way in contact with one another. (For instance the product of any two areas which overlap at two separated places will be of this sort.) This might be thought unwelcome[31].

It would be possible to avoid this consequence. For provided that our variables are *already* interpreted as ranging only over continuous or unseparated areas we could quite satisfactorily define

$$a \text{ is separated from } b$$

for

$$(\forall y)((\forall x)(x|a \ \& \ x|b \rightarrow x|y) \rightarrow (y|a \vee y|b)).$$

We could then so restrict the existential postulates that a and b are required to have a sum only if they are not separated (in the above sense), and more generally we could require all the F's to have a sum only if[32]

$$(\exists y)((\forall z)((\forall x)(Fx \rightarrow z|x) \rightarrow y|z) \ \& \ (\forall x)(Fx \rightarrow \sim x|y)).$$

Then, if we began without any 'separated' areas, none could be generated by the existential postulates. But the resulting system would be quite unlike P as formulated. For one thing, it would clearly be necessary to add extra axioms, for under the intended interpretation it must for example be true that

$$a \subset b \ \& \ b \not\subset a \rightarrow (\exists x)(x|a \ \& \ x \subset b \ \& \ \sim a \text{ is separated from } x)$$

---

[31] In the case of two separated *areas* intuition may be helped if we imagine them formed into one continuous figure by a *line* joining them. However this manoeuvre cannot be repeated with two separated *lines* (each parts of a longer line).

[32] If this were the sole condition on the existence of sums then the set-theoretical interpretation would look very dubious. For example, by taking Fx as $a \in x$ we could deduce the existence of a universal set.

but this could not be proved from an existential postulate that was restricted in the way suggested. More importantly, several very fundamental theorems of P would have to be abandoned, for instance the associativity of the sum. For if, as is obviously possible, there are three areas a, b, c such that a is touching b and b is touching c but a is separated from c, then on the intended interpretation we shall have

$$E!(a + (c + b))$$

but also

$$\sim E!((a + c) + b)$$

because indeed

$$\sim E!(a + c).$$

What emerges, then, is that although we *could* so arrange matters that no separated sums are required to exist, if we did we should definitely lose our analogy with the relation which holds between the members of one class and the members of another when the first class is a subclass of the second. (Clearly there is no sense in which one can distinguish between those discrete classes that are 'separated' and those which are 'in contact', and with classes there is no possible interpretation for the failure of the associativity of the sum.) However, this would make the system quite useless for our purpose, since our purpose is precisely to exploit the analogy between parts and subclasses. In that case it may well be asked why we do not adopt the apparently harmless assumption that the sum of any two elements always exists, for this presumably is the case with the sum (i.e. union) of any two classes. Since, as we have just seen, we must anyway allow disconnected sums, no new objection could be levelled against us on that score, and it is not clear what other objections there might be. I am entirely sympathetic with this line of thought. So far as concerns the formal development of our system it will turn out to make no difference whether or not we make this assumption (which I shall call the assumption that all finite sums exist), and therefore I shall not adopt it formally. However, I see no harm in adding it as a further and informal requirement on intuitively satisfactory part–whole relations, for we shall need some informal requirement in any case.

To see the necessity for this, we must observe that even

though it may perhaps be granted that all reasonable part–whole relations do fit the axioms for P, still the converse is evidently not true. One important class of relations which fit the axioms for P but which might not ordinarily be thought of as straightforward part–whole relations consists of relations formed by restricting an ordinary part–whole relation so that only certain of the parts of an object are allowed to count. Examples are '. . . is a vertical cross-section of . . .' and '. . . is a horizontal cross-section of . . .'. As we have seen (p. 106) it is just such relations that we shall actually require for numerical comparisons such as '. . . is twice as high as . . .' or '. . . is twice as long as . . .', so we may gladly class these as being themselves part–whole relations. But of course any number of quite non-standard interpretations will be possible, namely interpretations in which what is counted as the *sum* of several objects is not something that we recognize as being *made up out of* those objects. As there is no way in which a restriction to 'standard' interpretations can be put more formally, there seems nothing for it but to leave this attempt to analyse the notion of a part–whole relation incomplete. From a purely formal point of view we may permit as a part–whole relation any relation that satisfies our axioms P1 and P2, for this is all we need for the ensuing deductions. Informally, however, we shall have to add that any intuitively satisfactory part–whole relation must also be such that the sums defined in terms of it are reasonably viewed as *made up out of* their summands, and we may further add that all such *finite* sums should exist. Although this latter 'informal' condition could of course be formally incorporated into the axioms themselves, we shall not do so since (as I say) it will not be needed for the ensuing deductions, and anyway the first informal condition must remain informal.

One last detail must be filled in before we return to the topic of numerical comparison. Since the axioms for P are now to be thought of as schemata which might be satisfied by various distinct relations, it will be best to indicate this by writing a relation-variable in place of the apparently unambiguous $\subset$, and I accordingly adopt $\subset_\varphi$ as the official notation. This is intended as a type-neutral variable. It ranges over all dyadic homogeneous relations of any type which satisfy the axioms for P, at least when the variables in those axioms are restricted to range over the entities in the field of that relation—a restriction intended but left tacit in all the formulae of this section. To spell this out in detail let us revert once more to my usual type-neutral symbolism. For any dyadic homogeneous relation $\Re$ we abbreviate

$$\alpha \in C(\Re) \quad \text{for} \quad (\exists \beta)(\alpha \Re \beta \lor \beta \Re \alpha)$$

and we define discreteness for $\Re$ by putting

$$\alpha |_{\Re} \beta \quad \text{for} \quad \sim(\exists \gamma)(\gamma \in C(\Re) \,\&\, \gamma \Re \alpha \,\&\, \gamma \Re \beta).$$

Then $\Re$ satisfies P1 if and only if

(i) $(\forall \alpha \beta)(\alpha \in C(\Re) \,\&\, \beta \in C(\Re)$
$\rightarrow (\alpha \Re \beta \leftrightarrow (\forall \gamma)(\gamma \in C(\Re) \,\&\, \gamma |_{\Re} \beta \rightarrow \gamma |_{\Re} \alpha)))$

and $\Re$ satisfies P2 if and only if

(ii) $(\forall \Phi)(((\exists \alpha)(\alpha \in C(\Re) \,\&\, \Phi \alpha) \,\&\, (\exists \beta)(\beta \in C(\Re) \,\&\, (\forall \alpha)(\alpha \in C(\Re)$
$\&\, \Phi \alpha \rightarrow \alpha \Re \beta))) \rightarrow (\exists \beta)(\beta \in C(\Re) \,\&\, (\forall \gamma)(\gamma \in C(\Re)$
$\rightarrow (\beta |_{\Re} \gamma \leftrightarrow (\forall \alpha)(\alpha \in C(\Re) \,\&\, \Phi \alpha \rightarrow \alpha |_{\Re} \gamma))))).$

We then abbreviate (using PW for part–whole)

$$PW(\Re)$$

for the conjunction of (i) and (ii) above[33]. The official notation is therefore introduced by regarding any formula

$$(\forall \phi)(\mathscr{A}(\subset_{\varphi}))$$

as short for the corresponding formula

$$(\forall \Re)(PW(\Re) \rightarrow \mathscr{A}(\Re))$$

Thus $\subset_{\varphi}$ is an alternative notation for $\Re$ where $PW(\Re)$ is given. We similarly use $\alpha |_{\varphi} \beta$ and $\phi \alpha$ as alternative notations for $\alpha |_{\Re} \beta$ and $\alpha \in C(\Re)$, where $PW(\Re)$ is given. Further, such signs

---

[33] It is easily seen that most of the clauses restricting variables to the field of $\Re$ are in fact superfluous. Without affecting the definition of $PW(\Re)$ we can drop this clause from the definition of $\alpha |_{\Re} \beta$, we can drop all these clauses except the first from the statement of P1, and we can drop them all from the statement of P2. Further, P1 can be simplified to read

$$(\forall \alpha \beta)(\alpha \Re \beta \leftrightarrow (\exists \gamma)(\gamma \Re \alpha) \,\&\, (\forall \gamma)(\gamma |_{\Re} \beta \rightarrow \gamma |_{\Re} \alpha)).$$

as $+$, $\cdot$, $-$, defined in terms of $\subset_\varphi$, are officially understood as written with the same subscript $+_\varphi$, $\cdot_\varphi$, $-_\varphi$. Finally, in all formulae containing signs with the subscript ϕ the variables occurring in association with those subscripted signs are understood to be restricted to the entities α such that $\phi\alpha$.

Since the official notation is, of course, quite unnecessarily prolix for most purposes, I shall in practice continue to omit the subscript ϕ where it does not affect the argument and continue to leave tacit the restriction to the field of $\subset_\varphi$.

5. Axioms for Numerical Comparison

In §3 I argued that if a comparative relation is to support fundamental numerical comparisons, then those comparisons must be explicable in terms of parts, on the general pattern

a is n times as ϕ as b if and only if a contains and is exhausted by n mutually discrete parts, each as ϕ as b.

Now it clearly holds for any satisfactory numerical comparison that if a and b are each n times as ϕ as the same thing then they must be as ϕ as one another, and a natural way of securing this implication in fundamental cases is therefore by laying down the following thesis. If a and b each divide exhaustively into several discrete parts and if the parts of a are as ϕ as corresponding parts of b, then a is itself as ϕ as b. For simplicity let us begin with a division into just two discrete parts, where (omitting subscripts for brevity) the thesis can be stated as:

$$(\forall xyzw)(a = x + y \ \& \ x|y \ \& \ b = z + w \ \& \ z|w$$
$$\rightarrow (x \approx z \ \& \ y \approx w \rightarrow a \approx b)).$$

To express this more neatly we may notice that an arbitrary division of a into two discrete parts may always be represented by using the product $a \cdot x$ and the difference $a - x$ for arbitrary x, so the thesis could be stated more briefly as

ADD $(\forall xy)(a \cdot x \approx b \cdot y \ \& \ a - x \approx b - y \rightarrow a \approx b)$.

I shall call this thesis the elementary principle of addition.

Roughly speaking, the principle of addition can be seen as telling us that the magnitude (e.g. length) of a thing is some function of the magnitudes (lengths) of its parts, because it says that when the parts are equal in magnitude so are the wholes. But it does not tell us *which* function this is. To provide information on this point we may observe that, where fundamental numerical comparisons are in order, if some proper part of a is as ϕ as b then a must be more ϕ than b, and the converse holds also. Hence

$$a \succ b \leftrightarrow (\exists x)(x \subset a \ \& \ x \neq a \ \& \ x \approx b)$$

or, in the same idiom as ADD above,

COMP $a \succ b \leftrightarrow (\exists x)(a \cdot x \neq a \ \& \ a \cdot x \approx b).$

I shall call this thesis the principle of comparison. Let us now consider these two principles as a possible pair of axioms from which, together with axioms P1, P2 for a part–whole relation and the axioms C1, C2 for a comparative relation, one might try to build the theory of (fundamental) numerical comparison.

We may usefully begin by defining 'a is at least n times as ϕ as b', which I abbreviate (still omitting subscripts) to $a \succcurlyeq_n b$. Evidently this will be defined to mean 'a contains at least n mutually discrete parts, each of which is as ϕ as b'. For clarity let us temporally write

$$a \mathrm{D} b \quad \text{for} \quad a|b.$$

Then the definition can be put in the form

$$a \succcurlyeq_n b \quad \text{for} \quad (\exists n_{\mathrm{D}} x)(x \subset a \ \& \ x \approx b)$$

Since discreteness is a counting relation (p. 9) we have

$$a \succcurlyeq_1 b \leftrightarrow (\exists x)(x \subset a \ \& \ x \approx b)$$

$$a \succcurlyeq_{n'} b \leftrightarrow (\exists x)(x \subset a \ \& \ x \approx b \ \& \ (\exists n_{\mathrm{D}} y)(y \subset a \ \& \ y \approx b \ \& \ y|x))$$

whence it follows at once that

$$a \succcurlyeq_1 b \leftrightarrow a \succcurlyeq b$$

$$a \succcurlyeq_{n'} b \leftrightarrow (\exists x)(a \cdot x \approx b \ \& \ a - x \succcurlyeq_n b).$$

These recursive equivalences, together with our principles ADD and COMP, will allow us to prove many of the expected theses concerning the relation $\succcurlyeq_n$. We may also introduce other relations of numerical comparison in terms of $\geqslant_n$, for example defining

$$a \leqslant_n b \quad \text{for} \quad (\forall x)(x \geqslant_n b \to x \geqslant a)$$

$$a \approx_n b \quad \text{for} \quad a \geqslant_n b \,\&\, a \leqslant_n b$$

$$a \succ_n b \quad \text{for} \quad a \geqslant_n b \,\&\, a \nleqslant_n b$$

$$a \prec_n b \quad \text{for} \quad a \ngeqslant_n b \,\&\, a \leqslant_n b$$

and again several of the expected theses will be forthcoming. However, several will not, for there is one crucial thesis concerning the relation $\geqslant_n$ which we cannot yet prove, namely Archimedes' postulate:

ARCH $(\exists n)(a \prec_n b)$.

One might initially suspect that the reason why we cannot prove this postulate is that it very clearly requires that only finite wholes be taken into consideration, which is a rather noticeable restriction. For example, one might well hold that time has always existed and so be willing to say that the time that has elapsed up to now is infinitely greater than, say, one hour. In that case one will presumably hold that for each natural number n the time elapsed up to now is at least n times as great as one hour, and so will reject Archimedes' postulate in this application. I shall return shortly to the question whether it is reasonable to rule out infinite wholes, but first we should observe that this point cannot plausibly be said to be the reason why Archimedes' postulate is not deducible from our present basis. For that basis *already* contains a restriction to finite wholes. It is an immediate consequence of COMP that

$$b \subset a \,\&\, b \neq a \to a \succ b$$

which is the Euclidean axiom 'The whole is greater than the (proper) part'. But it has long been noticed that this Euclidean axiom does not hold for infinite wholes[34].

Let us consider, then, how we might try to deduce Archimedes' postulate from the Euclidean axiom. The general idea of the proof is this. Suppose that $(\forall n)(a \geqslant_n b)$. Then we can

[34] See e.g. Russell 1901.

mark off on a a progression of mutually discrete parts a_1, a_2, a_3, . . ., all of which are equal to b:

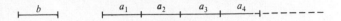

The sum of all these parts will exist, since they are all parts of a, and will itself be a part of a (and possibly the whole of a). But this sum will not obey the Euclidean axiom. For consider now the sum of the parts a_2, a_3, a_4, . . ., i.e. the first sum without a_1. This second sum is clearly a *proper* part of the first, but it will also be *equal* to the first, since the two sums can be matched term for term. That is, for each i we have $a_{i+1} \approx a_i$, and the second sum is the sum of the a_{i+1} for each i while the first sum is the sum of the a_i for each i. Therefore, in accordance with the thought underlying our principle of addition, the two sums should be equal. This clearly contradicts the Euclidean axiom, and hence if the Euclidean axiom holds so also must Archimedes' postulate.

Now this seems to me to be an intuitively acceptable proof, but it cannot be carried through from our present basis. One reason is quite evident. The principle of addition as formulated concerns only the sums of *two* discrete parts. By repeated uses of the principle we may extend it to cover any finite number of parts, but what we are concerned with here is a sum of infinitely many parts. So what is needed is a more inclusive formulation of the principle which will apply to sums of *any* number of mutually discrete parts (though in fact we shall not need, in the present system, to consider sums of more than countably many parts). To formulate our more inclusive version we shall have to be more explicit about the idea of *matching* two sums term for term. Let us first say that a relation is a *matching* relation if and only if it is one-to-one, and any terms related by the relation are equal. Thus

$$Matching\ (R)$$

for

$$(\forall xyzw)(xRy\ \&\ zRw \rightarrow (x = z \leftrightarrow y = w))\ \&\ (\forall xy)(xRy \rightarrow x \approx y).$$

Then we may say that a relation *matches* the things that are F with the things that are G if and only if it is a matching relation and the things that are F form its domain and the things that are G form its counterdomain, i.e.

$$R \text{ matches } \{x:Fx\} \text{ with } \{x:Gx\}$$

for

$$Matching(R) \ \& \ (\forall x)(Fx \leftrightarrow (\exists y)(xRy)) \ \& \ (\forall y)(Gy \leftrightarrow (\exists x)(xRy))$$

and finally we may usefully abbreviate

$Disc\Sigma\{x:Fx\}$ for $\ E!\Sigma\{x:Fx\} \ \& \ (\forall xy)(Fx \ \& \ Fy \ \& \ x \neq y \rightarrow x|y).$

The principle we require can now be formulated as

$$
\begin{aligned}
\text{ADD+} \ \ &Disc\Sigma\{x:Fx\} \ \& \ Disc\Sigma\{x:Gx\} \\
&\& \ R \text{ matches}\{x:Fx\} \text{ with } \{x:Gx\} \\
&\rightarrow \Sigma\{x:Fx\} \approx \Sigma\{x:Gx\}.
\end{aligned}
$$

I shall call this thesis ADD+, the *extended* principle of addition, in contrast to ADD, the *elementary* principle. Clearly ADD+ entails ADD as a special case.

The question now arises of whether or not, with ADD+ in place of ADD, we may complete our proof of Archimedes' postulate. Here the answer is that we may do so if (but only if) we may invoke our principle for constructing progressions, discussed in Chapter 1, §4. For the initial step of the proof just sketched was to claim that if $(\forall n)(a \succcurlyeq_n b)$ then we may mark off on a a progression of mutually discrete parts $a_1, a_2, a_3, \ldots,$ each equal to b. It seems obvious that this can in fact be done, for since $a \succcurlyeq_1 b$ we can clearly select a part a_1 of a which is equal to b (for the existence of such a part is given by COMP). Further, since $a \succcurlyeq_2 b$, there is certainly *some* part of a which is twice as great as b, and from this it follows that for *every* part a_1 of a equal to b there must be another part of a discrete from it and equal to b. Therefore, given a_1 we can select a_2. Similarly, given a_1 and a_2 we can select a_3, since $a \succcurlyeq_3 b$, and so on. But intuitively speaking, infinitely many selections are needed here, each depending on the previous selections, and that is why we must appeal to the version of the axiom of choice that allows for the construction of progressions. Granting this appeal, the argument may now be completed without further difficulty, and so Archimedes' postulate is in this way deducible from ADD+ and COMP.

A further and important principle that is deducible in a similar way from the same basis is the principle of least upper bounds, i.e. the principle that any non-empty set of things which has an upper bound has a least upper bound. That is

L.U.B. $(\exists x)(Fx)$ & $(\exists y)(\forall x)(Fx \rightarrow x \leqslant y)$

$\rightarrow (\exists y)(\forall z)(y \leqslant z \leftrightarrow (\forall x)(Fx \rightarrow x \leqslant z))$.

I shall sketch the proof of this principle in two stages. First consider the special case in which our non-empty set forms a *progression* under the relation $\prec$, i.e. the case where for some function $\mathbf{f}$ from numbers to objects and for some element a

$$(\forall n)\big(\mathbf{f}(n) \prec \mathbf{f}(n')\big) \ \& \ (\forall n)(\mathbf{f}(n) \leqslant a).$$

The general strategy of the proof is to show that in this case we can mark off on a a new progression of mutually discrete parts $a_1, a_2, a_3, \ldots$, so defined that the sum of all the parts a_i up to a_n is equal to the nth member of the original progression under $\prec$. Then we shall argue that the sum of *all* these parts a_i exists (since it is a part of a) and is itself a least upper bound for all the members of the original progression. For convenience, let us write

$$\sum_{i \leqslant n} \{a_i\}$$

for the sum of all the terms $a_1, a_2, \ldots, a_n$ up to and including a_n, and

$$\sum_{\forall i} \{a_i\}$$

for the sum of all the terms altogether. Then the progression of mutually discrete parts a_i is to be chosen so that

$$(\forall n)\big(\mathbf{f}(n) \approx \sum_{i \leqslant n} \{a_i\}\big)$$

and we have to show that

$$(\forall z)\Big(\sum_{\forall i} \{a_i\} \leqslant z \leftrightarrow (\forall n)(\mathbf{f}(n) \leqslant z)\Big).$$

Clearly we must begin by once more invoking our principle for constructing progressions. First, since $\mathbf{f}(1) \leqslant a$, we can certainly select a part a_1 of a equal to $\mathbf{f}(1)$. Next, since $\mathbf{f}(1) \prec \mathbf{f}(2) \leqslant a$, there is certainly some part of a equal to $\mathbf{f}(2)$

and greater than $\mathbf{f}(1)$, from which it follows that given any part a_1 of a equal to $\mathbf{f}(1)$ there exists another part a_2 of a discrete from a_1 and such that $a_1 + a_2$ is equal to $\mathbf{f}(2)$. Therefore, given a_1 we may select a_2. Similarly, given a_1 and a_2 we may select a_3 discrete from $a_1 + a_2$ and such that $a_1 + a_2 + a_3$ is equal to $\mathbf{f}(3)$, and so on. In this way we obtain the desired progression, and may move on to consider its sum.

It is easy to see that the sum we have constructed is an upper bound of the progression $\mathbf{f}$. For we have by construction that

$$(\forall n)\left(\mathbf{f}(n) \approx \sum_{i \leqslant n} \{a_i\}\right)$$

and of course

$$(\forall n)\left(\sum_{i \leqslant n} \{a_i\} \subset \sum_{\forall i} \{a_i\}\right)$$

whence

$$(\forall n)\left(\mathbf{f}(n) \leqslant \sum_{\forall i} \{a_i\}\right).$$

To show that this sum is a *least* upper bound we notice first that our original argument can be generalized to show that for *any* element z if

$$(\forall n)(\mathbf{f}(n) \leqslant z)$$

then there exists a progression z_1, z_2, z_3, ... of mutually discrete parts of z satisfying the condition

$$(\forall n)\left(\mathbf{f}(n) \approx \sum_{i \leqslant n} \{z_i\}\right)$$

and the sum of this whole progression is of course a part of z:

$$\sum_{\forall i} \{z_i\} \subset z.$$

Further, by the extended principle of addition, the sums of all such progressions of mutually discrete parts are equal to one another and hence are equal to the sum of our progression of mutually discrete parts of a

$$\sum_{\forall i} \{z_i\} \approx \sum_{\forall i} \{a_i\}$$

whence we deduce

$$\sum_{\forall i} \{a_i\} \leqslant z.$$

Conditionalizing this argument we have shown that

$$(\forall z)\Big((\forall n)(\mathbf{f}(n) \leqslant z) \rightarrow \underset{\forall i}{\Sigma} \, \{a_i\} \leqslant z\Big)$$

and hence that our sum of all the parts a_i is indeed a least upper bound. This completes the argument, and L.U.B. is now established at least for this special case.

Turning now to consider the principle more generally, we may dismiss as trivial the case in which our non-empty set has a greatest member, for in that case it is obvious that the greatest member will itself be a least upper bound of the set. Therefore let us now assume

(i) $(\exists x)(Fx)$

(ii) $(\forall x)(Fx \rightarrow x \leqslant a)$

(iii) $(\forall x)(Fx \rightarrow (\exists y)(Fy \,\&\, y \succ x))$.

The general strategy is to show that in this case we may select from among our non-empty set a progression under $\prec$ which has precisely the same upper bounds as the whole set. If this can be done, we have evidently reduced this more general case to the special case just treated, and the proof will be completed. Further, all that is necessary to ensure that our progression has the same upper bounds as the whole set is to see that the progression is so constructed that every member of the original set is less than some member of the selected progression, and this can indeed be achieved by making repeated use of Archimedes' postulate, which is now available to us.

First we use the postulate to establish, from assumptions (i)–(iii), that

$$(\forall x)(\exists y)(Fy \,\&\, (\forall zw)(Fz \,\&\, w \subset z \,\&\, w \neq z \,\&\, w \approx y \rightarrow x \succ z - w)).$$

What this says is that, given anything x, there will be a member of our set such that the difference between it and any greater member is less than x—that is, intuitively speaking, given an item x of any length (for example) we can find a member of the set which is within that length of the desired least upper bound. The next thing to show, then, is that we may choose our standard x as small as we please, so from the same assumptions we show that

$$(\forall n)(\exists x)(a \succ_n x).$$

Putting these last two theses together, we evidently have

$$(\forall n)(\exists y)(Fy \;\&\; (\forall zw)(Fz \;\&\; w \subset z \;\&\; w \neq z \;\&\; w \approx y \to a \succ_n z - w).$$

We may now invoke (for the last time) our principle for constructing progressions. For if we abbreviate the last thesis to

$$(\forall n)(\exists y)(\mathscr{A}(y,\, n))$$

it is a simple matter to show that

$$(\exists x)(\mathscr{A}(x,\, 1)) \;\&\; (\forall x)(\forall n)(\mathscr{A}(x,\, n) \to (\exists y)(x \prec y \;\&\; \mathscr{A}(y,\, n')).$$

This enables us to construct the desired progression, selected from the original set, whose members approach nearer and nearer to the desired least upper bound without limit. All that remains to be shown is that every member of the original set is less than some member of the progression, and this results quite straightforwardly from a further application of Archimedes' postulate. The principle of least upper bounds is thus now established in full generality.

Let us briefly review the discussion so far. We began with two elementary principles, of addition and of comparison, namely

ADD $(\forall xy)(a \cdot x \approx b \cdot y \;\&\; a - x \approx b - y \to a \approx b)$

COMP $a \succ b \leftrightarrow (\exists x)(a \cdot x \neq a \;\&\; a \cdot x \approx b).$

These two principles proved inadequate as a basis for our theory, so we were led to extend the principle of addition to cover infinite sums, i.e. to

ADD+ $Disc\Sigma \{x : Fx\} \;\&\; Disc\Sigma \{x : Gx\}$

         $\&\; R\; matches\; \{x : Fx\}\; with\; \{x : Gx\}$

         $\to \Sigma \{x : Fx\} \approx \Sigma \{x : Gx\}.$

Allowing appeals to our principle for constructing progressions, from ADD+ and COMP we were able to deduce first Archimedes' postulate

ARCH $(\exists n)(a \prec_n b)$

and finally the general principle of least upper bounds

L.U.B. $(\exists x)(Fx) \;\&\; (\exists y)(\forall x)(Fx \to x \leqslant y)$

         $\to (\exists y)(\forall z)(y \leqslant z \leftrightarrow (\forall x)(Fx \to x \leqslant z)).$

As will appear more fully hereafter (pp. 244–8), the availability of L.U.B. shows that we have now reached an adequate basis for our theory, so (granted the principle for constructing progressions) we could satisfactorily take ADD+ and COMP as the sole axioms. Alternatively we could begin with L.U.B. as an axiom, and this would allow some simplification of the remaining axioms. So far as I can see (though I may be mistaken here) L.U.B. does not actually entail ADD+, but it will always allow us to bypass ADD+ (and the associated appeals to the principle for constructing progressions) for we can always use least upper bounds instead of infinite sums. If infinite sums are not required, then the underlying logic of parts could obviously be simplified as well. For example, Archimedes' postulate can be proved directly from L.U.B. with (of course) the help of the elementary principles ADD and COMP but without invoking any infinite sums and therefore without appealing to ADD+ or to the principle for constructing progressions. The proof is as follows.

Assume, for *reductio ad absurdum*, that

(i) $(\forall n)(a \succcurlyeq_n b)$.

It is then obvious that

$$(\forall n)(\forall x)(x \approx_n b \rightarrow x \leqslant a)$$

which is to say that there exists an upper bound to the terms x such that $(\exists n)(x \approx_n b)$. Hence by L.U.B. there exists a least upper bound, say c, such that

(ii) $(\forall z)(c \leqslant z \leftrightarrow (\forall n)(\forall x)(x \approx_n b \rightarrow x \leqslant z))$.

Now since $c \leqslant c$ it follows that

$$(\forall n)(\forall x)(x \approx_n b \rightarrow x \leqslant c).$$

Therefore

$$(\forall n)(\exists x)(x \approx_n b) \rightarrow (\forall n)(\exists x)(x \approx_n b \,\&\, x \leqslant c)$$

and hence

$$(\forall n)(\exists x)(x \approx_n b) \rightarrow (\forall n)(c \succcurlyeq_n b).$$

However, it is easily seen from (i) that

$$(\forall n)(\exists x)(x \approx_n b)$$

so we therefore deduce

(iii) $(\forall n)(c \succcurlyeq_n b)$.

From this it follows at once that

$$c \succ b$$

and therefore there must be a division of c, say into $c \cdot d$ and $c - d$, such that

(iv) $c \cdot d \approx b \ \& \ c - d \prec c$.

Applying (ii) to the second conjunct here we infer

$$(\exists n)(\exists x)(x \approx_n b \ \& \ x \succ c - d).$$

Therefore

$$(\exists n)(c - d \prec_n b)$$

So, adding the first conjunct of (iv) again, we have

(v) $(\exists n)(c \cdot d \approx b \ \& \ c - d \prec_n b)$.

But from (v) it follows that

(vi) $(\exists n)(c \prec_{n+1} b)$

in direct contradiction to (iii). Thus we obtain our *reductio*, and thereby a proof of Archimedes' postulate, without any of the complicated argument that was previously needed.

An axiomatization which takes the principle of least upper bounds as fundamental has therefore much to recommend it from the point of view of formal simplicity, and in the next section I shall develop this approach a little further. However, I think it is also illuminating to see how the principle of least upper bounds can be derived by elaborating the connection between (fundamental) numerical comparisons and parts, so for the rest of the present section I shall return to the approach based on ADD+ and COMP.

A point that demands some discussion is the limitation to finite wholes, so first let us consider an axiomatization which is free from this limitation. Now no such limitation is implied by ADD or ADD+, and with our present axioms the limitation lies wholly in the principle of comparison, which we may consider in two halves:

COMP (i) $(\exists x)(x \subset a \ \& \ x \neq a \ \& \ x \approx b) \rightarrow a \succ b$

COMP (ii) $a \succ b \rightarrow (\exists x)(x \subset a \ \& \ x \neq a \ \& \ x \approx b)$.

COMP (i), as we have remarked, directly implies the Euclidean axiom of finity:

FIN[35] $\quad b \subset a \ \& \ b \neq a \rightarrow a \succ b.$

However, if COMP (i) is simply dropped, COMP (ii) will not be strong enough for our purposes, and it will be best to replace it with a revised principle of comparison:

COMP* $\quad a \leqslant b \leftrightarrow (\exists x)(a \approx x \ \& \ x \subset b).$

This revised principle is clearly acceptable for infinite as well as for finite wholes; it is stronger than COMP (ii), and it and FIN together are equivalent to the original COMP. Therefore if FIN is to be dropped, the first step is to replace COMP by the weaker COMP*.

On the other hand the principle of addition will need strengthening. Our original version could be phrased as

$$(\exists x)(a \cdot c \approx b \cdot x \ \& \ a - c \approx b - x) \rightarrow (a \cdot c) + (a - c) \approx b.$$

By applying COMP to this we could previously deduce its converse

$$(a \cdot c) + (a - c) \approx b \rightarrow (\exists x)(a \cdot c \approx b \cdot x \ \& \ a - c \approx b - x)$$

but the deduction depends upon the assumption of finity, and so will not be available if FIN is dropped. However, the converse itself seems equally acceptable for infinite wholes, and so the next step would seem to be to replace the original ADD by a biconditional version

$$(a \cdot c) + (a - c) \approx b \leftrightarrow (\exists x)(a \cdot c \approx b \cdot x \ \& \ a - c \approx b - x).$$

Now, however, we may notice a tempting analogy between the biconditional COMP* and this further biconditional. The analogy is clearer if we weaken the latter to

ADD* $(a \cdot c) + (a - c) \leqslant b \leftrightarrow (\exists xy)(a \cdot c \approx x \cdot y \ \& \ a - c \approx x - y \ \& \ x \subset b).$

ADD* is clearly deducible from the stronger biconditional and COMP* together; it is strong enough to imply the conditional

---

[35] We may conveniently *define* a finite whole as one which is greater than each of its proper parts:

$$Fin(a) \quad \text{for} \quad (\forall x)(x \subset a \ \& \ x \approx a \rightarrow x = a).$$

ADD without any trouble, and it also turns out that it implies
the earlier biconditional version of ADD, granted some further
argument to follow shortly. So ADD* would suffice for our
purposes. What is tempting, however, is that ADD* very clearly
says the same thing of a division of a into two discrete parts as
COMP* says of the (trivial) division of a into one (discrete)
part. Therefore if we generalize to *any* division of a into discrete
parts we shall have an extended principle of addition which
yields not only ADD* but also COMP* as a special case. The
required generalization is evidently

$$(\text{ADD}+)^* \quad Disc\Sigma\{x:Fx\} \rightarrow (\Sigma\{x:Fx\} \leqslant b \leftrightarrow$$
$$(\exists R)(\exists G)(Disc\Sigma\{x:Gx\} \ \& \ \Sigma\{x:Gx\} \subset b$$
$$\& \ R \ matches \ \{x:Fx\} \ with \ \{x:Gx\})).$$

This will suffice as the sole axiom if we do not wish to limit our
consideration to finite wholes.

Let us briefly consider the way in which the theory of
numerical comparison could be developed from this basis. First,
we do not of course expect ARCH itself as a theorem if infinite
wholes are permitted, but we may well expect that the argument previously given for ARCH (pp. 132–4) would remain intact
until its last step,. That is, we expect that if $(\forall n)(a \succcurlyeq_n b)$ then
we can mark out on a a progression of mutually discrete parts,
each equal to b. This progression will then be equal to a proper
part of itself (*viz.* the progression without its first member),
and so a also will be equal to a proper part of itself. Hence we
may expect to have in place of ARCH the weaker result

ARCH* $(\forall n)(a \succcurlyeq_n b \rightarrow (\exists x)(a \cdot x \approx b \ \& \ a - x \approx a).$

The expectation proves to be fulfilled, though the original
argument requires a little modification. This is because in the
infinite case $(\forall n)(a \succcurlyeq_n b)$ does not imply $a \succ b$ but is quite
consistent with $a \approx b$. (Indeed, if $\approx$ is taken as 'as many as'
and a is a denumerable class, it will follow that $(\forall n)(a \approx_n a)$,
for any denumerable class may be divided into any number of
discrete but still denumerable subclasses.) So in effect we proceed by considering separately the case in which $a \succ b$ and
the case in which $a \approx b$, though it proves technically more convenient to distinguish these as the case in which $(\forall n)(a \succ_n b)$

and the case in which $(\exists n)(a \approx_n b)$. In the first the original argument goes through without much difficulty, and in the second we argue simply that if

$$(\forall n)(a \succcurlyeq_n b) \ \& \ (\exists n)(a \approx_n b)$$

then for some n

$$a \succcurlyeq_{n'} b \ \& \ a \approx_n b.$$

$$\therefore \quad (\exists x)(a \cdot x \approx b \ \& \ a - x \succcurlyeq_n b \ \& \ a \approx_n b)$$

$$\therefore \quad (\exists x)(a \cdot x \approx b \ \& \ a - x \succcurlyeq a).$$

and therefore

$$(\exists x)(a \cdot x \approx b \ \& \ a - x \approx a)$$

which is the desired conclusion.

Although ARCH* is thus available, so far as I can see (and again I may be mistaken here) L.U.B. is not forthcoming without further assumptions. We *could*, however, obtain L.U.B. if we allowed ourselves to make use of transfinite ordinals and to appeal to the full axiom of choice. Very briefly, the main strategy of the argument would be as follows. First it is easy to check that we still have available the argument given earlier (pp. 135–7) for the special case of L.U.B. where our non-empty set forms a progression under $\prec$. Next we observe that this argument may be generalized to cover the case where our non-empty set is *well ordered* by $\prec$, so that its members can each be designated as $\mathbf{f}(\alpha)$ for some function $\mathbf{f}$ defined on ordinals and such that for all ordinals α and β we have $\mathbf{f}(\alpha) \leqslant \mathbf{f}(\beta) \rightarrow \alpha \leqslant \beta$. (This generalization of the argument depends on the fact that where λ is a limit ordinal then, in the notation used previously,

$$\sum_{\alpha < \lambda} \{a_\alpha\} \leqslant \mathbf{f}(\lambda).$$

The argument actually given establishes this for $\lambda = \omega$.) Next we notice that if our non-empty set is well-ordered by any relation at all then we can select from it a subset which is well-ordered by $\prec$ and which has the same upper bounds as the whole set. (The selected subset may be defined by stipulating that, for each ordinal α, $\mathbf{f}(\alpha)$ is to be the earliest element x in the original well-ordering such that $(\forall \beta)(\beta < \alpha \rightarrow \mathbf{f}(\beta) \prec x)$.)

Then finally we argue that according to the axiom of choice every non-empty set is well-ordered by some relation or other, and hence from every non-empty set we can select a subset well-ordered by $\prec$ which has the same upper bounds, and hence that every non-empty set which has an upper bound has a least upper bound.

In this proof the use made of the full axiom of choice seems to be crucial, and so far as I can see there is no way of replacing it by a weaker assumption. Therefore I regard it as doubtful whether L.U.B. should be accepted where infinite wholes are concerned, but I am also inclined to think that this doubt is not too worrying. For in those places where one does naturally appeal to L.U.B. (e.g. pp. 173–7, 238) one would anyway not be concerned with infinite wholes, and besides the fundamentals of the theory of numerical comparison are certainly forthcoming without it. Indeed the next phase of development invokes only COMP*, ADD*, and ARCH*.

From these three principles we establish first an important 'subtraction' thesis:

$$(\forall xy)(a \cdot x \not\approx a \rightarrow (a \approx b \;\&\; a \cdot x \approx b \cdot y \rightarrow a - x \approx b - y)).$$

This incidentally enables us to recover from ADD* our original and stronger biconditional

$$(a \cdot c) + (a - c) \approx b \leftrightarrow (\exists x)(a \cdot c \approx b \cdot x \;\&\; a - c \approx b - x).$$

More importantly it enables us to prove a neat pair of recursive equivalences for the relation $\approx_n$ to match those for $\succcurlyeq_n$, namely

$$a \approx_1 b \leftrightarrow a \approx b$$

$$a \approx_{n'} b \leftrightarrow (\exists x)(a \cdot x \approx b \;\&\; a - x \approx_n b)$$

and another neat pair of recursive equivalences for the relation $\preccurlyeq_n$, namely

$$a \preccurlyeq_1 b \leftrightarrow a \preccurlyeq b$$

$$a \preccurlyeq_{n'} b \leftrightarrow (\forall x)(a \cdot x \approx b \;\&\; \text{E!}(a - x) \rightarrow a - x \preccurlyeq_n b).$$

Further results are now quite straightforward.

From all this we may conclude that there is certainly no *need* to restrict our attention to finite wholes in order to develop a satisfactory theory of numerical comparison. All the basic results, as we have seen, can be obtained from COMP*, ADD*, and ARCH*. Hence (allowing appeal to our principle for constructing progressions) they can be obtained from COMP*, ADD*, and ADD+, and it is tempting to replace these three axioms by the single axiom (ADD+)*:

$$Disc\Sigma\,\{x\!:\!Fx\} \to (\Sigma\,\{x\!:\!Fx\} \leqslant b \leftrightarrow (\exists R)(\exists G)(Disc\Sigma\,\{x\!:\!Gx\}$$
$$\&\; \Sigma\,\{x\!:\!Gx\} \subset b \;\&\; R\; matches\; \{x\!:\!Fx\}\; with\; \{x\!:\!Gx\})).$$

(ADD+)* is stronger than we need, in two ways. First, the implication from right to left need never be applied to sums that are more than denumerable, though it is hard to see this as a real extravagance because it is quite unclear how the implication this way round might hold for denumerable sums but not for sums of higher cardinality. Second, the implication from left to right need never be applied to sums of more than *two* members, so the axiom does have a great deal of superfluous power when taken in this direction. In fact given COMP* and ADD* one can of course prove that the implication from left to right holds for sums with finitely many members, and given also ADD+ (applied to denumerable sums) and our principle for constructing progressions one can go on to prove that the implication holds also for denumerable sums. (This is a corollary of the proof sketched on pp. 135–8.) Further, no *finite* whole can be a sum of more than denumerably many *discrete* parts, so I think that there need be no doubt of the acceptability of our single axiom in the *finite* case. However, in the infinite case these assurances seem not to be available, and one may reasonably say that the superfluous strength of (ADD+)* introduces a quite unnecessary doubt into the theory. So far as I can see, any reasonable comparative relation which satisfies COMP*, ADD*, and ADD+ will also satisfy (ADD+)*, but this conjecture is without much foundation, and my *guess* is that if it were to prove false we would rather abandon (ADD+)* than conclude that COMP*, ADD*, and ADD+ were after all inadequate. After all, the main temptation to (ADD+)* is only that it allows us to simplify the axiomatic basis to a single (organic) axiom.

This temptation could lead us to yet stronger assumptions in the same direction. The assumption we are considering at present is that if $a \leqslant b$, and a can be represented as a sum of any number of discrete elements, then *there exists* a relation which matches that sum term by term with some discrete parts of *b*. Now, it may be asked, why should we restrict this assumption to cases where *a* is represented as a *discrete* sum? The restriction could be lifted if we were prepared to assume the existence of a relation which matched *every* part of *a* with a part of *b* and preserved discreteness, thus yielding the neater axiom

$$a \leqslant b \leftrightarrow (\exists x)(x \subset b \ \& \ (\exists R)(R \ matches \ \{y : y \subset a\} \ with \ \{y : y \subset x\}$$
$$\& \ (\forall xyzw)(xRy \ \& \ zRw \rightarrow (x|z \rightarrow y|w)))).$$

This axiom would yield COMP* and ADD* easily enough, and it would yield ADD+, at least for denumerable sums, provided we may use our principle for constructing progressions. So it would be quite sufficient. However, it is much stronger than necessary, and may well seem pretty dubious. After all, there are several cases where we are sure that $a \leqslant b$ but there seems to be no prospect of ever *specifying* the relation which this axiom requires to exist.

The moral to draw from this discussion is evidently not that infinite wholes are unmanageable, and if, as seems to be the case, the standard accounts imply that a quantity which permits infinite wholes cannot be measured, then those standard accounts are simply in error. In fact it is quite easy to lay down axioms which are evidently acceptable and which provide for all undoubted theses. However, there are difficulties which arise when we consider strengthening these axioms to form a more complete theory, since various extra assumptions look tempting, but as our grasp of infinity is not very secure our untutored intuitions give us no very firm guidance on the matter. Clearly further investigation is required. Of course there has already been a great deal of further investigation with one particular comparative relation, the relation 'as many as', but this is a somewhat special case. The quantity concerned here (say 'numerosity') is a discrete quantity, whereas one would wish to frame a general theory which covered both discrete and continuous quantities impartially.

Despite all this I shall not, in this book, pursue the question of infinite wholes any further. From the point of view of the present chapter we are after all only interested in numerical comparison for the sake of the light that it sheds on rational (and real) numbers. Looking a little further back, these investigations in turn sprang from an enquiry into natural numbers and the way they are related to rational (and real) numbers. It may reasonably be doubted whether a detailed pursuit of infinite wholes would forward this original enquiry, especially as infinite wholes must evidently be put on one side when we come to consider ratios and proportions, and that is where rational (and real) numbers become relevant. Obviously there can be no ratio between the finite and the infinite, or at any rate none that can be expressed by a rational number, while between two infinite wholes either there will again be no ratio or there will be every possible ratio. So the theory of ratios applied to infinite wholes is of no interest at all, and for the purposes of that theory we lose nothing and gain greatly in simplicity if we do exclude infinite wholes. Accordingly I do exclude them from now on. Since, as I have remarked, the axiom (ADD+)* is not at all doubtful in this case, I shall adopt it as my sole remaining axiom.

This axiom 'associates' a part–whole relation and a comparative relation, and to see its purport clearly we shall have to remember the suppressed subscripts. I have said that from a purely formal point of view any relation $\subset_\varphi$ which satisfies our axioms P1 and P2 may be permitted to count as a part–whole relation, though there are also informal conditions to be met by any intuitively acceptable part–whole relation. Similarly we may say that from a formal point of view any relation $\succcurlyeq_\psi$ may be permitted to count as a comparative relation provided it satisfies our axioms C1 and C2[36]. Now we add that a comparative relation $\succcurlyeq_\psi$ and a part–whole relation $\subset_\varphi$ are *associated* if and only if (i) the fields of the two relations coincide, i.e. $(\forall x)(\phi x \leftrightarrow \psi x)$, and (ii) the two relations together satisfy the axiom (ADD+)*, at least when its variables are restricted to elements in the field of the relations. Strictly, then, (ADD+)* should be formulated with appropriate subscripts ϕ and ψ throughout (and the clause $x = y$ in the definition of 'matching'

---

[36] That is, $\succcurlyeq_\psi$ is a type-neutral relation variable ranging over all dyadic homogeneous relations that satisfy C1 and C2, whatever their type. The details may be filled in as on pp. 128–30.

should be written as $x \subset_\varphi y \,\&\, y \subset_\varphi x$). I also say that a comparative relation gives rise to fundamental numerical comparisons if and only if it is associated in this way with some part–whole relation. The axioms, that is to say, are intended to constitute a definition or analysis of the concept of fundamental numerical comparison.

As such, they have one awkward feature, for according to these formal definitions there is nothing to prevent a comparative relation being associated with several distinct part–whole relations and in such a way that the numerical comparisons based on distinct part–whole relations do not coincide. However, when we add our informal conditions on part–whole relations we see that this latter possibility is ruled out. For informally we are requiring that any two elements in the field of a part–whole relation will have a sum (as defined in terms of that relation) and this sum will actually be composed of the two summands and nothing else. Hence two elements in the field of several part–whole relations will have a sum according to each of those relations, and all these sums will coincide. Supposing, then, that $\subset_\varphi$ and $\subset_\psi$ are each part–whole relations associated with the same comparative relation $\succcurlyeq$ (I omit the subscript in this case) and writing the respective sums as $x +_\varphi y$ and $x +_\psi y$, our informal conditions certainly yield the consequence

    (i) $(\forall xy)(x +_\varphi y \approx x +_\psi y)$.

Now this further implies

    (ii) $(\forall xy)(x|_\varphi y \leftrightarrow x|_\psi y)$

at least in the finite case, which is sufficient for our purposes. For in that case if

    (iii) $a|_\varphi b$

then

    $a +_\varphi b \succ a$

so by assumption (i)

    $a +_\psi b \succ a$

and hence for some element c

    $a +_\psi b \approx a +_\psi c \,\&\, a|_\psi c \,\&\, c \subset_\psi a +_\psi b.$

$\therefore$ $a +_\psi b \approx a +_\psi c \,\&\, a|_\psi c \,\&\, c \subset_\psi b$

∴(iv) $a +_\psi b \approx a +_\psi c$ & $a|_\psi c$ & $c \leqslant b$.

Now since $a|_\psi c$, by repetition of the same argument from (iii) to (iv) with ϕ and ψ interchanged, we deduce that for some element d

(v) $a +_\varphi c \approx a +_\varphi d$ & $a|_\varphi d$ & $d \leqslant c$.

From the first clauses of (iv) and (v) we have

$$a +_\psi b \approx a +_\psi c \text{ & } a +_\varphi c \approx a +_\varphi d$$

and therefore, in view of assumption (i),

$$a +_\varphi b \approx a +_\varphi d$$

but this, together with (iii), yields

$$b \leqslant d$$

and so, in view of the last clause of (v),

(vi) $b \leqslant c$.

Putting (iv) and (vi) together we have

$$a +_\psi b \approx a +_\psi c \text{ & } a|_\psi c \text{ & } b \approx c$$

from which there follows the desired conclusion

(vii) $a|_\psi b$.

The argument from (iii) to (vii) evidently establishes (ii), and from (i) and (ii) together we of course have

$$(\forall xy)(a \approx x +_\varphi y \text{ & } x|_\varphi y \leftrightarrow a \approx x +_\psi y \text{ & } x|_\psi y).$$

Since all numerical comparisons are defined in terms of sums of discrete elements, it clearly follows that all numerical comparisons based on $\subset_\varphi$ and $\subset_\psi$ must coincide.

Formally speaking, then, a relation of fundamental numerical comparison '. . . is at least n times as ϕ as . . .' is built up from a numerical quantifier 'there are at least n . . .', a comparative relation '. . . is at least as ϕ as . . .', *and* some part–whole relation. However, informally, if we confine ourselves to intuitively acceptable part–whole relations, then it will not matter which particular part–whole relation is chosen.

But cf. p.157
165–6

6. The Traditional Axioms: Magnitudes

The sole axiom (ADD+)* of the previous section was designed
to apply to infinite wholes no less than finite ones, but henceforth I impose also a requirement of finity. Therefore we shall
proceed with (ADD+)* and FIN as axioms, from which, as
remarked earlier, we can easily obtain the original axiomatization based on ADD+ and COMP. The salient steps in our first
construction of the system were that we began with two
elementary principles:

ADD $(\forall xy)(a \cdot x \approx b \cdot y \ \& \ a - x \approx b - y \to a \approx b)$

COMP $a \succ b \leftrightarrow (\exists x)(a \cdot x \neq a \ \& \ a \cdot x \approx b)$.

Strengthening ADD to ADD+ we then deduced

ARCH $(\exists n)(a \prec_n b)$

and hence, with a further application of ADD+,

L.U.B. $(\exists x)(Fx) \ \& \ (\exists y)(\forall x)(Fx \to x \leqslant y)$

$\qquad \to (\exists y)(\forall z)(y \leqslant z \to (\forall x)(Fx \to x \leqslant z))$.

Let us now compare this approach with what I earlier (pp.
102–3) called an orthodox set of postulates for 'fundamental
measurement', though I went on to raise an objection to that
label. The postulates, which I shall now call instead the traditional postulates for extensive quantities, are

(1) $E \,!\, a \oplus b$

(2) $a \oplus b \approx b \oplus a$

(3) $(a \oplus b) \oplus c \approx a \oplus (b \oplus c)$

(4) $a \approx b \to a \oplus c \approx b \oplus c$

(5) $a \succ b \leftrightarrow (\exists x)(a \approx b \oplus x)$.

To these should be added Archimedes' postulate, which is
usually given in the form

$$(\exists n)(a \prec n \circ b)$$

where $n \circ b$ is in effect introduced as

$$\overbrace{b \oplus b \oplus b \oplus \ldots \oplus b}^{n \text{ times}}.$$

One may also add L.U.B. (unchanged), though it is often omitted[37]. For the moment, however, let us concentrate mainly on (1)–(5).

Assuming the existence of all finite sums, the first three of these postulates are of course available when $\oplus$ is interpreted as $+$, but the last two are not. In place of (4) we have rather

ADD′ $a|c \ \& \ b|c \rightarrow (a \approx b \rightarrow a + c \approx b + c)$

and in place of (5) we can deduce[38]

COMP′ $a \succ b \leftrightarrow (\exists x)(x|b \ \& \ a \approx b + x)$.

This evidently reflects the point I made earlier, that on the traditional accounts it seems that $a \oplus b$ exists only when a and b are discrete. Yet, as we see from the way in which Archimedes' postulate is framed, this is not in fact in harmony with the traditional accounts, for in these accounts terms such as $a \oplus a$ figure crucially as paradigms for what is twice as great as a.

We shall approach the spirit of the traditional accounts more nearly if we introduce what I shall call (following Russell) *indefinite* descriptions. The idea here is that a formula containing an expression $a \oplus b$ shall be interpreted as saying something about *some element or other*, it does not matter which, which is a sum of two discrete parts, one equal to a and the other equal to b. We shall resolve to use these indefinite descriptions only as 'terms' to comparative relations, so it will suffice to introduce them as 'terms' to the relation $\succcurlyeq$, regarding this as the primitive comparative relation. By analogy with our earlier introduction of definite descriptions (pp. 114–5) we may therefore stipulate that where η, θ are any *quasi-terms* (i.e. any variables or definite descriptions or indefinite descriptions), then where ζ is any quasi-term

---

[37] My limited reading would support the generalization that L.U.B. is omitted by philosophers of science (e.g. Nagel 1932, Suppes 1951, Ellis 1966) and included by mathematicians or philosophers of mathematics (e.g. Forder 1927, pp. 19–24, and Professor D. S. Scott in an unpublished treatment he has been good enough to show me). I discuss the question in the Appendix to Chapter 3.

[38] In fact COMP′ is weaker than COMP. Assuming ADD′ and the existence of all finite sums a principle equal to COMP in deductive power is

$$a \succ b \leftrightarrow (\exists x)(x \subset a \ \& \ x|b \ \& \ a \approx b + x).$$

$$\eta \oplus \theta \succcurlyeq \zeta$$

abbreviates

$$(\exists xy)(x \approx \eta \;\&\; y \approx \theta \;\&\; x|y \;\&\; x+y \succcurlyeq \zeta)$$

and where ξ is any genuine term

$$\xi \succcurlyeq \eta \oplus \theta$$

abbreviates

$$(\exists xy)(x \approx \eta \;\&\; y \approx \theta \;\&\; x|y \;\&\; \xi \succcurlyeq x+y).$$

Further, where θ is any indefinite description (more will be introduced shortly), it will be convenient to use

$$\mathrm{E}!\theta$$

to abbreviate

$$(\exists x)(x \approx \theta)$$

so in the present case we have

$$\mathrm{E}!a \oplus b \leftrightarrow (\exists xyz)(y \approx a \;\&\; z \approx b \;\&\; y|z \;\&\; x \approx y+z).$$

It is then easily seen that so long as our indefinite descriptions are confined to contexts definable in terms of $\succcurlyeq$ they will behave just as definite descriptions do, for we may easily show that

$$\mathrm{E}!a \oplus b \rightarrow ((\forall x)(\mathscr{A}(x)) \rightarrow \mathscr{A}(a \oplus b))$$

wherever $\mathscr{A}(a \oplus b)$ results from $\mathscr{A}(x)$ upon replacing all free occurrences of x in $\mathscr{A}(x)$ by occurrences of $a \oplus b$.

Now even if we do assume the existence of all finite sums, still we shall not have $\mathrm{E}!a \oplus b$ as a thesis. (It is largely because of this that it makes no difference to the theory of numerical comparison whether or not we make this assumption.) In particular we have no warrant to assume that $\mathrm{E}!a \oplus a$, because we have no warrant to assume that there is anything greater than a let alone twice as great as a. Indeed in some cases the assumption is surely false. For example, according to current scientific theorizing the universe is finite in volume, and therefore it *cannot* be true of *every* sum of physical objects that there is another sum of physical objects as large as it and discrete from it; there must be a bound on the size of (sums of) physical objects. On the other hand it would be absurd to suppose that because of this we cannot measure the size of physical objects fundamentally and cannot employ fundamental numerical comparisons in this respect. It follows that the assumption $E!a \oplus b$ *must* be unnecessary to our purpose, though so far as one can see it is taken to be necessary in the standard accounts.

Therefore, on our way of introducing $\oplus$, we find that postulate (1) is simply not forthcoming. To take its place we shall instead require an axiom-schema to the effect that if any complex quasi-term exists then so do its components; i.e. we shall require, in the case of the quasi-operator $\oplus$, that for any quasi-terms ζ, θ

(1') $E!\zeta \oplus \theta \rightarrow E!\zeta \ \& \ E!\theta.$

(Analogous results will then be deducible for other quasi-operators.) Of course the validity of this schema is an immediate consequence of our definition of the quasi-operator $\oplus$. The remaining postulates are available provided that we add the appropriate existential conditions, namely

(2') $E!a \oplus b \rightarrow (a \oplus b \approx b \oplus a)$

(3') $E!(a \oplus b) \oplus c \rightarrow ((a \oplus b) \oplus c \approx a \oplus (b \oplus c))$

(4') $E!a \oplus c \rightarrow (a \approx b \rightarrow (a \oplus c \approx b \oplus c))$

(5') $a \succ b \leftrightarrow (\exists x)(a \approx b \oplus x).$

These are deducible from ADD and COMP whether or not we assume the existence of all finite sums, but their existential conditions cannot be dispensed with.

To continue with this style of developing the theory, we may usefully introduce some more indefinite descriptions. Thus $a \ominus b$ may be used to stand for some element or other, it does not matter which, which is a difference between something equal to a and one of its parts equal to b. We could define (for η, θ, ζ, ξ, as before)

$\eta \ominus \theta \succcurlyeq \zeta$ for $(\exists xy)(x \approx \eta \ \& \ y \approx \theta \ \& \ y \subset x \ \& \ x - y \succcurlyeq \zeta)$

$\xi \succcurlyeq \eta \ominus \theta$ for $(\exists xy)(x \approx \eta \ \& \ y \approx \theta \ \& \ y \subset x \ \& \ \xi \succcurlyeq x - y).$

However, with the quasi-operator $\oplus$ already to hand it is simple enough to define $\ominus$ directly in terms of $\oplus$. For example we could put

$\zeta \approx \eta \ominus \theta$ for $\zeta \oplus \theta \approx \eta$

and then define other contexts for $\ominus$ in terms of this one in an obvious way, e.g.[39]

$$\zeta \succcurlyeq \eta \ominus \theta \quad \text{for} \quad (\exists x)(\zeta \succcurlyeq x \ \& \ x \approx \eta \ominus \theta)$$

$$\eta \ominus \theta \succcurlyeq \xi \quad \text{for} \quad (\exists x)(x \succcurlyeq \xi \ \& \ x \approx \eta \ominus \theta).$$

The two schemes of definition evidently coincide, and on either scheme it is easy to see that

$$E!\eta \ominus \theta \leftrightarrow \eta \succ \theta.$$

We may also introduce '$n{\circ}a$' as an indefinite description. In terms of our original system the definition is easily given thus:

$$n{\circ}\theta \succcurlyeq \zeta \quad \text{for} \quad (\exists xy)(x \approx \theta \ \& \ y \approx_n x \ \& \ y \succcurlyeq \zeta)$$

$$\xi \succcurlyeq n{\circ}\theta \quad \text{for} \quad (\exists xy)(x \approx \theta \ \& \ y \approx_n x \ \& \ \xi \succcurlyeq y).$$

From these definitions we easily prove the recursive equivalences

$$(\forall x)(x \approx 1{\circ}a \leftrightarrow x \approx a)$$

$$(\forall x)(x \approx n'{\circ}a \leftrightarrow x \approx (n{\circ}a) \oplus a)$$

from which we can see that for each n we shall indeed have

$$n{\circ}a \approx \overbrace{a \oplus a \oplus \ldots \oplus a}^{n \text{ times}}$$

provided that $E!n{\circ}a$. However, this latter is not a thesis, for in fact

$$E!n{\circ}a \leftrightarrow (\exists x)(x \approx_n a).$$

If on the other hand we wish to introduce $n{\circ}a$ directly in terms of $\oplus$, the numerical quantifiers which underlie our definition of $\approx_n$ will be of no help to us, and the simplest expedient is to turn instead to numerical powers of relations (Chapter 1, §2). We shall apply this idea to relations of the form $\ldots \succcurlyeq \theta \oplus \ldots$ so for clarity let us write

$$\zeta(\mathbf{S}_\theta)\eta \quad \text{for} \quad \zeta \succcurlyeq \theta \oplus \eta.$$

---

[39] We have unconditionally

$$\eta \ominus \theta \succcurlyeq \zeta \leftrightarrow \eta \succcurlyeq \zeta \oplus \theta$$

but the counterpart to this thesis carries existential conditions

$$E!\zeta \oplus \theta \rightarrow (\zeta \succcurlyeq \eta \ominus \theta \rightarrow \zeta \oplus \theta \succcurlyeq \eta)$$

$$E!\eta \ominus \theta \rightarrow (\zeta \oplus \theta \succcurlyeq \eta \rightarrow \zeta \succcurlyeq \eta \ominus \theta).$$

Then the required definition can be given as

$$\zeta \geqslant 1 \circ \theta \quad \text{for} \quad \zeta \geqslant \theta$$

$$\zeta \geqslant n' \circ \theta \quad \text{for} \quad \zeta (\mathbf{S}_\theta)^n \theta.$$

With an analogous definition for $n \circ \theta \geqslant \zeta$, the same recursive equivalences will evidently be forthcoming. Alternatively we could of course simply regard the recursive equivalences themselves as a pair of extra axioms, stipulating in our formation rules that $n \circ a$ is to be an expression of the same syntactic category as $a \oplus a$, or we could employ any of the well-known methods for replacing a pair of recursive equivalences by an explicit definition.

On any of these approaches we shall have

$$a \succ_n b \leftrightarrow a \succ n \circ b$$

$$a \approx_n b \leftrightarrow a \approx n \circ b$$

but the same does *not* hold for $\prec$. This is because our formula $a \prec_n b$ introduced no terms (or quasi-terms) other than a and b, and in effect it holds whenever a and b are in the field of the relation $\geqslant$ but a is not at least n times as great as b. Hence this formula does not imply that anything *is* at least n times as great as b. On the other hand $a \prec n \circ b$ is defined as a variant on $n \circ b \geqslant a$ & $a \not\geqslant n \circ b$, and the first conjunct here does imply $\mathrm{E}!n \circ b$. In fact we have

$$\mathrm{E}!n \circ b \to (a \prec_n b \leftrightarrow a \prec n \circ b)$$

$$\sim \mathrm{E}!n \circ b \to (a \prec_n b \ \& \ \sim a \prec n \circ b).$$

Since we do not assume $\mathrm{E}!n \circ b$, it is clear that in this reformulation Archimedes' postulate should be phrased in the more cautious form

$$(\exists n)(a \not\geqslant n \circ b).$$

L.U.B. is of course available unchanged, and (as we saw, pp. 139–40) when it is included Archimedes' postulate becomes superfluous.

In this 'reformulation' what we have been doing is deriving

from the original system a further set of postulates from which the notion of a *part* has been entirely eliminated. In the new set of postulates we may take just $\geqslant$, $\oplus$, and n° as primitive, or we may reduce these to just $\geqslant$ and $\oplus$, or we may if we wish reduce further to just $\approx$ and $\oplus$. In place of axioms C1 and C2 for comparative relations we should then have axioms assuring us that $\approx$ is an equivalence relation, and using (5') as a *definition* of $>$ we could thence recover C1 at once. The remaining properties of comparative relations would then follow if we added the further postulates

$$a \not\approx a \oplus b$$

$$a \not\approx b \to (\exists x)(a \oplus x \approx b) \lor (\exists x)(a \approx b \oplus x).$$

A further reduction of primitive notions would be to replace the dyadic relation $\approx$ and the quasi-operator $\oplus$ by a single triadic relation which holds between a, b, c when and only when $a \approx b \oplus c$. This formulation is of some technical convenience, partly because an extensive quantity is then represented by just one relation, and partly because with quasi-terms once more introduced by definition the axiom-schema (1') becomes superfluous, so that the theory is finitely axiomatizable. The simplest plan is then to regard the symbol ϕ, which officially occurs as subscripts in our formulae, as itself a variable for such an underlying triadic relation, taken as primitive. The postulates are then[40]

(a) $\sim\phi(a, a, b)$

(b) $\phi(a, b, c) \to \phi(a, c, b)$

(c) $(\exists x)(\phi(a, b, x) \,\&\, \phi(x, c, d)) \to (\exists x)(\phi(a, d, x) \,\&\, \phi(x, c, b))$

(d) $(\exists x)(\phi(a, b, x)) \lor (\exists x)(\phi(b, a, x))$
$\lor (\forall xy)(\phi(x, a, y) \leftrightarrow (\phi(x, b, y))).$

To these we may add L.U.B., e.g. in the form

(e) $(\exists x)(Fx) \to ((\forall y)(\exists xw)(Fx \,\&\, \phi(x, y, w))$
$\lor (\exists z)(\forall y)((\exists w)(\phi(z, y, w)) \leftrightarrow (\exists xw)(Fx \,\&\, \phi(x, y, w)))).$

---

[40] For simplicity I again omit an explicit restriction of the variables to the field of ϕ. Strictly, a and b need to be so restricted in postulate (d).

Starting with these postulates we would define

$$a \approx b \quad \text{for} \quad (\forall xy)(\phi(x, a, y) \leftrightarrow \phi(x, b, y))$$

$$a \succ b \quad \text{for} \quad (\exists x)(\phi(a, b, x)).$$

The quasi-operator $\oplus$ could be introduced in the first place by abbreviating (for ζ, η, θ any quasi-terms)

$$\zeta \approx \eta \oplus \theta \quad \text{for} \quad (\exists xyz)(x \approx \zeta \ \& \ y \approx \eta \ \& \ z \approx \theta \ \& \ \phi(x, y, z))$$

and other contexts for $\oplus$ could be defined in terms of this one. It is easy to check that all our original postulates, including the axiom-schema (1'), are thereupon forthcoming.

However, these variations, though technically useful, are of no great philosophical importance. What is philosophically interesting is that we may derive, from our original postulates for part–whole relations and comparative relations, a new set of postulates which (as we shall see) suffices for the deduction of all properties of numerical comparative relations, but from which the notion of a part has been entirely eliminated. The new postulates are to all appearances just the postulates for extensive quantities laid down in the traditional approach, except that we do not assume E!$a \oplus b$, but the intended interpretation would seem to be somewhat different. We have introduced the expression $a \oplus b$ as an *indefinite* description, but we have seen that in the contexts available in the new postulates it behaves just as a definite description does. On the traditional approach $a \oplus b$ seems to be intended as a definite description right from the start, though so far we have been rather at a loss to explain just what it is a description of. Another related feature is that the relation $\approx$ was clearly a mere equivalence relation in the original system and did not licence substitution in all contexts (e.g. $a \approx b \rightarrow (\forall x)(a \subset x \leftrightarrow b \subset x)$ was certainly *not* a thesis). However, in the new postulates $\approx$ does behave just like an identity-relation. These two points together suggest an entirely new line of interpretation. Whereas previously we took the variables of the system to range over the various things in the field of some comparative relation, for example the things that have size or weight or duration, the new suggestion is evidently to take the variables as ranging over the sizes, weights, or durations themselves. In my earlier terminology

[margin note: But cf. p.149 165-6]

(pp. 89–90) the new suggestion is to take the variables as ranging over the degrees of some quantity, or what are often called magnitudes[41], rather than over the things that have or exemplify them. Under this interpretation it seems obvious that all our axioms remain equally acceptable, and it also seems quite wrong not to adopt the strong existential postulate

$$E\,!a \oplus b.$$

Whereas it is evidently rash to assume that every degree of a quantity is exemplified, it seems quite harmless to say that the degrees themselves exist whether or not they are exemplified. Indeed if we do want to hold that unexemplified sizes, for example, do not exist, how can we explain what is meant by the perfectly sensible claim that there are (there *exist*) some sizes which are not exemplified?

First it is worth pointing out that the currently popular view[42] which identifies a size with a class of objects having that size can be shown to be wrong by considering just this question, for on this view all sizes which are not exemplified must be identified with the null class, and therefore with each other, which is clearly not what is wanted. If sizes are countenanced at all we shall surely want to say that the size which a thing *would* have if it were a billion times larger than the sun, and the size which a thing *would* have if it were two billion times larger than the sun, must anyway be different sizes, even though it may turn out that neither of them is exemplified.

It was, I think, Frege who first proposed the identification of a magnitude with the class of things having that magnitude[43], but in his discussion, as in Russell's early discussion of magnitudes[44], the identification with classes is of secondary importance. The principal point that both are urging in their very different ways is that magnitudes should be accepted as abstract objects, and this view could be defended without making the mistake of identifying them with classes and so assigning them

---

[41] Terminology in this area is quite unsettled. I call size (for instance) a quantity, and the various different sizes different magnitudes of that quantity. The opposite way of speaking is equally—perhaps more—defensible. Surprisingly, Russell (1903, Chap. 19) calls the things that *have* sizes quantities.

[42] See for example Suppes 1951, p. 40.

[43] See Frege 1884, §§ 62–9 (on directions).

[44] Russell 1903, Chap. 19.

the wrong identity conditions. What seems crucial is just the claim that we should recognize an expression such as 'the size of . . .' as a functional expression which, when supplied with an appropriate argument, yields an abstract object—a size—as value. It is evidently in harmony with this claim if we go on to add that an expression of the sort 'the size an object *would* have if . . .' may also succeed in referring to a size, and perhaps a size that is not exemplified. Clearly the view *need* not be developed in such a way as to imply that all non-exemplified sizes are identical[45].

The most obvious difficulty with this proposal, as with almost all proposals about abstract objects, is that unless we put some restriction on the means for generating these objects we are liable to be led into contradictions like the Russell paradox. For suppose we claim that any equality-relation $\approx_\varphi$ generates a series of abstract objects, which we may call ϕ-nesses. Then for anything a in the field of such a relation there will be an abstract object referred to as 'the ϕ-ness of a', and the converse also holds. Now clearly we must have, as an axiom governing these abstract objects,

(A) The ϕ-ness of $a =$ the ϕ-ness of $b \leftrightarrow a \approx_\varphi b$.

This axiom gives the truth-conditions for identity-statements where each term is given as a magnitude of the same family, *viz.* a ϕ-ness. However, to say nothing of Frege's question whether Julius Caesar might turn out to be a ϕ-ness, we must surely provide more information than this. Could it turn out, for example, that some sizes were also weights, or that every size was a weight? Here the natural answer is 'no': sizes and weights, if admitted as abstract objects at all, are surely always distinct objects. On the other hand if the question is raised say, about largenesses and smallnesses, one would probably be inclined to the opposite conclusion. There seems to be only an irrelevant difference between the relations '. . . is as large as . . .' and '. . . is as small as . . .'. Anyway, however exactly the question is to be answered, it seems very reasonable to require that some answer be provided, and one would certainly expect the answer to take the form of a condition on the two relations

[45] Perhaps Frege would not have agreed. Compare Dummett 1973, pp. 502–4.

$\approx_\varphi$ and $\approx_\psi$, since ϕ-nesses and ψ-nesses are surely to be understood *via* our grasp of these relations. The condition might be, for example, that the two relations were necessarily equivalent, but whatever it is let us represent it schematically by $\mathscr{I}(\phi, \psi)$. Then we shall expect to have, as a thesis

B(i) The ϕ-ness of a = the ψ-ness of b → $\mathscr{I}(\phi, \psi)$

and we shall *also* expect to have, as a converse,

B(ii) $\mathscr{I}(\phi, \psi) \to (\forall x)(x = $ the ϕ-ness of $a \leftrightarrow x = $ the ψ-ness of a).

Now from B(i) and B(ii) we evidently deduce

The ϕ-ness of a = the ψ-ness of b →
the ϕ-ness of a = the ϕ-ness of b

and thence, by employing (A),

(C) The ϕ-ness of a = the ψ-ness of $b \to a \approx_\varphi b$.

Now we are already in danger of a contradiction, for theses (A) and (C) between them enable us to define a *univocal* relation, say the relation of HAVING, which holds between any object and each of its various magnitudes[46], i.e. we may write

b HAS the ϕ-ness of a

for

$(\exists\psi)($the ϕ-ness of $a = $ the ψ-ness of $b)$

and in view of (A) and (C) we at once deduce

$(\forall x)(x$ HAS the ϕ-ness of $a \leftrightarrow x \approx_\varphi a)$

whence

$(\exists y)(\forall x)(x$ HAS $y \leftrightarrow x \approx_\varphi a)$.

To obtain an unacceptable result from this we only have to agree further that the equivalence relation which holds between any objects x and y if and only if $Fx \leftrightarrow Fy$ is a proper case of an equality-relation $\approx_\varphi$. If this is granted we deduce

$(\exists y)(\forall x)(x$ HAS $y \leftrightarrow (Fx \leftrightarrow Fa))$

and therefore

$Fa \to (\exists y)(\forall x)(x$ HAS $y \leftrightarrow Fx)$

---

[46] It is interesting that Russell's early discussion seems to deny the existence of any such univocal relation despite the fact that in a footnote on the same page he explicitly provides one, *viz.* the $\in$ of class-membership (Russell 1903, § 157).

whence
$$(\exists x)(Fx) \rightarrow (\exists y)(\forall x)(x \text{ HAS } y \leftrightarrow Fx).$$

However, this is a thesis which (as we have already had occasion to notice, p. 123) is totally unacceptable, since it implies that only one object exists.

Of course the argument just sketched could be blocked at many points. In particular, though (A) seems to be quite fundamental to this conception of magnitudes, it may be possible to resist (B) or (C), and in fact this has been proposed. For example Dummett has suggested[47] that no harm will be done if we identify the supposed abstract object, the ϕ-ness of a, with some concrete object, perhaps the earliest object as ϕ as a in some well-ordering of all concrete objects. It is evidently a consequence of this proposal that, say, the size of a and the weight of a might turn out to be the same thing although the size and weight of b were quite distinct things. Therefore (B) would be rejected, and by a similar example (C) would be rejected. However, to adopt this somewhat surprising proposal is evidently to abandon the idea that sizes are abstract objects at all, and in so doing to fall prey to the earlier objection: on this proposal it would be impossible for there to be any sizes which were not exemplified.

An alternative escape would be to place restrictions on which equivalence-relations were to count as equality-relations $\approx_\varphi$ and in particular to reject the relevant relations of the form $Fx \leftrightarrow Fy$. Of course this might be done by requiring that the equality-relation (or rather, its associated comparative relation) was 'associated' with some part–whole relation in such a way as to satisfy our axiom (ADD+)*, but that seems altogether too strong a requirement. The axiom (ADD+)* is intended to apply only where the quantity concerned yields *fundamental* numerical comparisons, and it does not apply to density or temperature for example. However it seems reasonable to hold that if sizes are abstract objects then so also are densities and temperatures. A weaker requirement, which at least *might* hold for density and temperature (see pp. 166–7) is that the equality-relation satisfy our more traditional set of postulates framed in terms of $\oplus$ for some interpretation of this sign. But that is too

[47] Dummett 1973, pp. 535–9.

weak since the equivalence relations $Fx \leftrightarrow Fy$ all *do* satisfy this requirement[48]. No doubt it is possible to suggest other restrictions (see e.g. p. 101) but it should at least be clear that it will not be too easy to hit on one that is entirely satisfactory.

Nor is this at all surprising. It is after all the chief lesson of Russell's paradox that we cannot regard *every* predicate-expression as picking out an (abstract) object with identity conditions at least as exacting as those for classes[49], and the present argument shows that the position is not likely to be improved if we restrict attention to predicate-expressions of the form '... $\approx_\varphi a$'. Of course consistency can be restored if, as in the modern theory of types, we refuse to permit variables to range over both abstract and concrete objects. For example in the present case we would then disallow the inference from

$$(\forall x)(x \text{ HAS the } \phi\text{-ness of } a \leftrightarrow x \approx_\varphi a)$$

to

$$(\exists y)(\forall x)(x \text{ HAS } y \leftrightarrow x \approx_\varphi a)$$

on the ground that the ϕ-ness of a, though we are calling it an abstract object, cannot be within the range of any variable y which is also used to quantify over concrete objects. However, the only rationale for this refusal seems to be to acknowledge that it is only a convenient *façon de parler* to speak of the supposed abstract objects as *objects* at all, so let us now consider a reductive approach to magnitudes, by which they are not admitted as *objects*, so that all apparent *references* to magnitudes are eliminated.

In ordinary speech a magnitude, such as a size, is generally specified by means of a condition which something would have to satisfy to be of that size, just as a class is generally specified by means of a condition of membership. Very often a remark ostensibly about a size may be paraphrased by dropping the apparent reference to a size and instead speaking directly of the things that satisfy the condition. Thus 'the size of a is greater than the size of b' may obviously be paraphrased to 'whatever is as large as a is larger than whatever is as large as b', or more simply to 'a is larger than b'. The same move is

---

[48] We stipulate that $a \oplus b$ is to exist if and only if $\sim Fa$ & $\sim Fb$, and when it does exist it is to be anything c such that Fc.

[49] See Volume I, Chapter 2, §2.

available where the sizes in question are—or may be—unexemplified, for then they will be specified counterfactually and the counterfactual will reappear in the paraphrase. For example 'the size a thing would have if it were one billion times larger than the sun is not the same as the size a thing would have if it were two billion times larger than the sun' can evidently be paraphrased to 'if there were anything one billion times larger than the sun it would not be as large as anything that was two billion times larger than the sun'. Generally, where the sizes in question are explicitly specified in this way, there will not be much difficulty in finding a paraphrase which eliminates the apparent references to sizes, and the notation which appears to involve these references may always be treated as syncategorematic. (This is the way class-notation is used in the present book.) However, more difficulty may be expected when we are faced with a remark that speaks of sizes quite generally without specifying which.

For the most part this difficulty can be met by an analogue of Russell's 'no-class' theory: we replace a generalization apparently about all sizes by a generalization concerning all conditions which (as we would say) specify sizes. For example since for every size there exists an object x and a natural number n such that the size can be specified as the size of anything which is n times as large as x, it is quite adequate to render 'some sizes are not exemplified' by $(\exists x)(\exists n) \sim (\exists y)(y \approx_n x)$ (where $\approx$ means 'as large as'). But not all cases are quite as simple as this. For example let us take as a test case 'sizes do not have parts as physical objects do (and so cannot satisfy our first axiomatization) but they do satisfy the traditional postulates', and let us take the second half of this sentence first.

Here it is more convenient to trespass on the next chapter and to make use of the fact that for any object a every size can be specified as the size of anything which is N times as large as a for some (positive) *real* number N. Accordingly we shall consider the conditions '... $\approx_N a$' (yet to be defined) for all real numbers N and some fixed object a (and for $\approx$ as 'as large as'). However, to speak of these conditions in the general way that we require we shall need an abstraction notation, say $\lambda x : x \approx_N a$, which can conveniently be read as 'what is said of x when it is said that $x \approx_N a$'. It should be noted that this last expression is not plausibly regarded as an expression for an object, for what is said of x when it is said that $x \approx_N a$ is precisely that it is N times as large as a, but 'that it is N times as large as a' is an incomplete (unsaturated) expression, because of the dangling 'it'. So our abstracts are to be thought of as having the syntax of predicate-expressions, just as Russell's class-expressions are, but they cannot be so simply introduced because we shall want them to have tighter identity-conditions than classes. For illustration it will suffice to assume that the required identity-conditions can be expressed by a modal operator, and then, on the pattern of Russell's no-class theory, we could introduce our λ-abstracts generally for all contexts by defining

$$(-\lambda x : x \approx_N a-) \quad \text{for} \quad (\exists F)(\Box(\forall x)(Fx \leftrightarrow x \approx_N a) \& (-F-))$$

(with appropriate rules for scope) but it is simpler to stick to the one context we require, which is to introduce a relation $\approx$ holding between λ-abstracts. Thus we might define

$$(\lambda x : x \approx_N a) \approx (\lambda x : x \approx_M a) \quad \text{for} \quad \Box(\forall x)(x \approx_N a \leftrightarrow x \approx_M a).$$

The important point is just that, however the definition is arranged, we shall require the consequence

$$(\lambda x : x \approx_N a) \approx (\lambda x : x \approx_M a) \leftrightarrow N = M.$$

Then we further define $\oplus$ between λ-abstracts:

$$(\lambda x : x \approx_N a) \oplus (\lambda x : x \approx_M a) \quad \text{for} \quad (\lambda x : x \approx_{N+M} a)$$

and we can now verify the postulates. For example the strong existential postulate $\mathrm{E}!a \oplus b$ is, under this interpretation,

$$(\exists P)((\lambda x : x \approx_P a) \approx (\lambda x : x \approx_N a) \oplus (\lambda x : x \approx_M a))$$

and clearly this and all the other postulates can be established by elementary reasoning with real numbers. The general strategy is therefore to get the effect of speaking of all sizes by instead speaking of all conditions that specify them, and this is then accomplished in turn by speaking of all real numbers that may be embedded in those conditions. It may be objected that we have only succeeded in paraphrasing a quantification over one sort of abstract object, sizes, by means of another quantification over another sort of abstract object, real numbers, but even if this objection is granted some reduction has certainly been achieved. For whereas previously we appeared to have not only sizes but also weights, durations, and many other magnitudes to cope with, *all* these may now be 'eliminated' in a similar way in favour of real numbers. Anyway it is not clear that the objection should be granted, as we shall see when we come to discuss the topic of real numbers.

The type of reduction just sketched may be applied to very many cases where we naturally speak of sizes (or other magnitudes), but it does not apply to all. In particular it does not apply to our other test case, 'sizes do not have parts as physical objects do'. For if $\subset$ is a part–whole relation which can meaningfully be applied to physical objects it will not even be *grammatical* to apply that relation between our λ-abstracts, since they have the syntax of predicates. The best we can do on this style of analysis is to retreat to a metalanguage and point out that if '$a \subset b$' makes sense then '$\ldots \subset (\lambda x : x \approx_N a)$' does not. But it is not very plausible to claim that this grammatical truth adequately represents the original thesis. Now the original thesis may be treated as purely hypothetical, 'if any objects are sizes then they do not have parts as physical objects do' (and in this case it is presumably true), or it may be treated as partly existential and implying 'there are objects which are sizes' (in which case I take it to be at best doubtful). Either way it seems that an analysis will have to explain what is *meant* by saying of an object that it is a size, and I am afraid that I do not at present see how that is to be explained. In the forgoing pages we have mentioned several features which sizes (as abstract objects) must be supposed to possess if they exist at all—for example there are presumably as many sizes as there are real numbers, sizes obey the traditional postulates in terms of $\oplus$, sizes obey the principles (A) and (B) (which threaten a contradiction), there may well be unexemplified sizes, and so on. However, these points do not seem to yield any satisfactory paraphrase for 'x is a size', even though we are no longer requiring that a satisfactory paraphrase make this sometimes true. In that rather unsatisfactory state I am afraid that I must leave the topic of magnitudes.

One thing more should be said on the reformulation of our system in terms of $\oplus$. This more traditional formulation, which eliminates the notion of a part, in fact preserves all that is required for developing the theory of numerical comparison, and hereafter my deductions will depend only on the reformulated postulates. This gives the deductions a wider application which is very useful, since the natural numbers and the real numbers satisfy the more traditional postulates though the notion of a part is hardly applicable to them[50]. (At the same time, as we saw in §3, the way is open to more unexpected interpretations for quite familiar quantities such as length.) Anyway the subsequent definitions and theorems for numerical comparisons will apply perfectly well to such things as numbers, even though in this case the numerical comparisons are not *fundamental* numerical comparisons. The question may be raised, then, whether perhaps *all* numerical comparisons, derived as well as fundamental, can be seen as flowing from the traditional postulates. In one way this is certainly the case, for if ϕ-ness is *any* quantity in respect of which numerical comparisons are definable—fundamental or derived—then the various *degrees* of that quantity, the various ϕ-nesses, will certainly satisfy the postulates in just the (reductive) way in which we have explained that sizes do. However, the objects that have (some degree of) ϕ-ness still need not satisfy the postulates, for postulate (5) makes a stronger existential claim than need be satisfied in the case of derived numerical comparisons, namely

$$a > b \rightarrow (\exists x)(a \approx b \oplus x).$$

Consider for example the comparatively favourable case of relative velocities, where we can easily supply a sense for the 'addition' of such velocities in terms of the objects that have them, though this has nothing to do with parts in the normal sense. Suppose we already understand the locutions

a is moving from *b* as fast as *c* is moving from *d*

a is moving from *b* in the same direction as *c* is moving from *d*

which I shall abbreviate to

$$\langle a, b \rangle \approx \langle c, d \rangle$$

$$\langle a, b \rangle \parallel \langle c, d \rangle$$

---

[50] Note that we once again assume the type-neutrality of the postulates. The claim that numbers (or sizes) satisfy the postulates does not imply that numbers (or sizes) are objects.

then in quite the orthodox manner we may interpret

$$\langle a, b \rangle \oplus \langle c, d \rangle$$

as an indefinite description for a pair of separating objects $\langle x, y \rangle$ such that

$$(\exists z)(\langle a, b \rangle \approx \langle x, z \rangle \ \& \ \langle c, d \rangle \approx \langle z, y \rangle \ \& \ \langle x, z \rangle \| \langle z, y \rangle).$$

However, it is highly doubtful whether this entirely orthodox interpretation satisfies postulate (5), for that now requires that if a is moving from b faster than c is moving from d then there must actually exist a trio of objects x, y, z such that

$$\langle c, d \rangle \approx \langle x, z \rangle \ \& \ \langle x, z \rangle \| \langle z, y \rangle \ \& \ \langle a, b \rangle \approx \langle x, y \rangle$$

But there seems no reason to suppose that this existential claim is always satisfied. If we generalize (and speak somewhat roughly), postulate (5) requires that if there is an object with N degrees of ϕ-ness and another with M degrees of ϕ-ness (where $N > M$) then there must exist a third object with $N - M$ degrees of ϕ-ness which is suitably related to an object with M degrees to be 'added' to it in some appropriate sense of $\oplus$. However, in the case of *derived* numerical comparisons there is no reason to suppose that this existential claim is satisfied, whatever shifts we go to to introduce an appropriate sense of 'addition' for these cases. As we have seen, the claim may well be satisfied where we are dealing with, say, magnitudes or numbers which are not (concrete) objects, but in the case of (concrete) objects we can expect it to be satisfied only where *fundamental* numerical comparisons would be in order, i.e. where the new objects required to exist can always be taken as *parts* of the original objects.

7. Fractions, Ratios, and Rational Numbers

Setting aside the excursus in the previous section, where we allowed ourselves (prematurely) to make use of relations of the form $\ldots \approx_N \ldots$ for any (positive) real number N, we have so far given formal consideration only to numerical comparisons employing natural numbers. However, one of our main reasons for paying attention to numerical comparisons was that they provide the primary setting for rational numbers. In this section we shall give a formal introduction of this use of rational numbers.

As remarked earlier, rational numbers can occur as *fractions*, and we shall consider this first. If I give you, say, a quarter of an apple, then clearly I give you some *part* of an apple, and this part of the apple is a quarter as large as the whole apple, which is to say that the whole apple is four times as large as it. Therefore, for an appropriate part–whole relation and comparative relation, we might say

$$a \text{ is a quarter of } b \rightarrow a \subset b \ \& \ b \approx 4{\circ}a.$$

If instead I give you three-quarters of an apple, then the part I give you must be three times as large as anything that is one-quarter of the apple, so we might be led to say

$$a \text{ is three-quarters of } b \leftrightarrow a \subset b \ \& \ (\forall x)(b \approx 4{\circ}x \rightarrow a \approx 3{\circ}x).$$

On reflection, however, although this account works well enough for apples, it would not apply in all cases. For if we consider something that does not have quarters, e.g. a set of just two members, or just 99 members, there surely should be nothing that is three-quarters of it. On the present definition, however, *every* part (subset) is trivially three-quarters of it. So we must alter the definition to

$$a \text{ is three-quarters of } b \leftrightarrow a \subset b \ \& \ (\exists x)(b \approx 4{\circ}x \ \& \ a \approx 3{\circ}x)$$

and generally

$$a \text{ is } (n/m)^{\text{ths}} \text{ of } b \leftrightarrow a \subset b \ \& \ (\exists x)(a \approx n{\circ}x \ \& \ b \approx m{\circ}x).$$

Now in working out the arithmetical properties of fractions thus defined, it soom becomes apparent that the clause $a \subset b$ does no work except to ensure that if a is $(n/m)^{\text{ths}}$ of b then $n \leqslant m$. It is altogether more convenient, therefore, to introduce an intermediate definition

$$a \approx \frac{n}{m}{\circ}b \quad \text{for} \quad (\exists x)(a \approx n{\circ}x \ \& \ b \approx m{\circ}x)$$

which of course will apply whether or not $n \leqslant m$. (We may regard it as introducing vulgar, and not necessarily proper, fractions.) As is indicated by the notation, the definition may be thought of as introducing a quasi-term (or indefinite description), to be read as 'n m^{ths} of b', which stands for something or other—it does not matter what—which is n times as great as something or other—it does not matter what—such that b is m times as great as it. From the formal point of view the simplest plan is probably to introduce first the quasi-term 'an mth of b' which, in accordance with our previous method for quasi-terms (p. 151–2), may be done by defining

$$\frac{1}{m}\circ\theta \succcurlyeq \zeta \quad \text{for} \quad (\exists x)(\theta \approx m\circ x \,\&\, x \succcurlyeq \zeta).$$

$$\xi \succcurlyeq \frac{1}{m}\circ\theta \quad \text{for} \quad (\exists x)(\theta \approx m\circ x \,\&\, \xi \succcurlyeq x)$$

We then further abbreviate

$$\frac{n}{m}\circ\theta \quad \text{for} \quad n\circ\left(\frac{1}{m}\circ\theta\right)$$

and what was cited above as an 'intermediate' definition then falls out as a consequence.

To prevent misunderstanding it may be well to make some comments on this new quasi-term at once. First, we find as expected that $(1/1)\circ a$ is interchangeable with a, and hence that $(n/1)\circ a$ is interchangeable with $n\circ a$. We also find that just as the three quasi-terms

$$n\circ(m\circ a), \quad m\circ(n\circ a), \quad (n\cdot m)\circ a$$

are interchangeable, so also are[51]

$$\frac{1}{n}\circ\left(\frac{1}{m}\circ a\right), \quad \frac{1}{m}\circ\left(\frac{1}{n}\circ a\right), \quad \frac{1}{n\cdot m}\circ a.$$

However, this analogy does not extend very far. For example the quasi-terms

$$n\circ\frac{1}{m}\circ a, \quad \frac{1}{m}\circ n\circ a$$

are *not* always interchangeable. This is because $2\circ(\tfrac{1}{2}\circ a)$, for

---

[51] Since $(n\circ m)\circ a$ is not well formed, but would have to be understood as an attempt at $(n\cdot m)\circ a$, the brackets in $n\circ(m\circ a)$ are superfluous. The same applies to the brackets in such expressions as

$$n\circ\left(\frac{1}{m}\circ a\right) \quad \text{or} \quad \frac{1}{n}\circ\left(\frac{1}{m}\circ a\right)$$

so I henceforth omit them.

instance, will exist if and only if $\frac{1}{2}{\circ}a$ exists, while $\frac{1}{2}{\circ}(2{\circ}a)$ will exist if and only if $2{\circ}a$ exists, and these conditions are by no means equivalent[52]. However, if both do exist they will certainly be equal to one another and to a. Somewhat similarly, although we do indeed have, as an immediate consequence of the definition,

$$a \approx \frac{1}{n}{\circ}b \leftrightarrow n{\circ}a \approx b$$

it is *not* true that

$$a \succ \frac{1}{n}{\circ}b \leftrightarrow n{\circ}a \succ b$$

or that

$$a \prec \frac{1}{n}{\circ}b \leftrightarrow n{\circ}a \prec b.$$

In both cases the implication from right to left fails because the right-hand side does not imply $E!(1/n){\circ}b$, and in the first case the implication from left to right fails because the left-hand side does not imply $E!n{\circ}a$.

In fact it turns out that many of the theorems concerning fractions (as thus defined) depend on a hypothesis that two numbers are mutually prime (i.e. have no common factor greater than 1). The chief lemma that one needs is

$$Prime(n, m) \rightarrow ((\exists x)(x \approx n{\circ}a \,\&\, x \approx m{\circ}b)$$

$$\rightarrow (\exists x)(a \approx m{\circ}x \,\&\, b \approx n{\circ}x))$$

which is established by fairly straightforward reasoning about mutually prime numbers. With the aid of this lemma one can then show, for example, how fractions may be added and multiplied. We first establish the special cases[53]

$$Prime(m, q) \rightarrow (\forall xy)(x \approx \frac{n}{m}{\circ}y \oplus \frac{p}{q}{\circ}y \leftrightarrow x \approx \frac{n{\cdot}q + m{\cdot}p}{m{\cdot}q}{\circ}y)$$

[52] However, the case is different if n and m are mutually prime, as observed below.

[53] For brevity I have omitted an existential condition on the theorems for multiplication, namely $E!(p/q){\circ}y$ for the first and $E!(p{\cdot}k/q){\circ}y$ for the second. This condition is required only for the implication from right to left, and is anyway superfluous if either $p{\cdot}k \leqslant q$ or $m{\cdot}k \leqslant n$. Hence it does not affect the applications below.

$$Prime(m, p) \leftrightarrow (\forall xy)(x \approx \frac{n}{m} \circ \frac{p}{q} \circ y \leftrightarrow x \approx \frac{n \cdot p}{m \cdot q} \circ y).$$

Then it is not difficult to improve these results to

$$Prime(m, q) \rightarrow (\forall xy)(x \approx \frac{n}{m \cdot k} \circ y \oplus \frac{p}{q \cdot k} \circ y \leftrightarrow x \approx \frac{n \cdot q + m \cdot p}{m \cdot q \cdot k} \circ y)$$

$$Prime(m, p) \rightarrow (\forall xy)(x \approx \frac{n}{m \cdot k} \circ \frac{p \cdot k}{q} \circ y \leftrightarrow x \approx \frac{n \cdot p}{m \cdot q} \circ y).$$

These second theorems in effect cover all cases, for any two numbers n_1 and n_2 will of course have a highest common factor k such that

$$(\exists mp)(n_1 = m \cdot k \ \& \ n_2 = p \cdot k \ \& \ prime(m, p)).$$

Now according to the second thesis here, the relation of being (for example) three-quarters of two-fifths of is always equivalent (for every subscripted ϕ^{54}) to the relation of being three-tenths of. However, it would be quite natural to report this fact in slightly different language by saying 'three-quarters of two-fifths (of) *is the same fraction as* three-tenths (of)', which we might indeed write as

$$\frac{3}{4} \circ \frac{2}{5} = \frac{3}{10}.$$

Similarly for addition we might report the general truth that (for example) whatever is one-sixth of something added to three-quarters of it must also be eleven-twelfths of it, and *vice versa*, by saying 'one-sixth added to three-quarters (of) is the same fraction as eleven-twelfths (of)', and write this as

$$\frac{1}{6} \oplus \frac{3}{4} = \frac{11}{12}.$$

There could obviously be no objection to reporting the facts in this way if our apparently singular statements about fractions were understood to mean no more and no less than the generalizations from which they have been, as it were, abstracted.

---

[54] More strictly, both $\succcurlyeq_\phi$ and $\oplus$ need to be subscripted. For a formal elucidation of the expression 'for every subscripted ϕ' (which I use constantly) see pp. 190–1.

However, this line of thought might easily lead to some further developments.

First we may notice that there are clear analogies between the way in which our signs $\oplus$ and $\circ$ function between fractions and the way in which $+$ and $\cdot$ function between natural numbers. For example it is easy to show that

$$(\forall xy)(x \approx (n{\circ}y) \oplus (m{\circ}y)) \leftrightarrow x \approx (n+m){\circ}y)$$

$$(\forall xy)(x \approx n{\circ}(m{\circ}y) \leftrightarrow x \approx (n \cdot m){\circ}y)$$

and so, 'abstracting' from these generalizations in the same way as before we may therefore write

$$n \oplus m = n + m$$

$$n{\circ}m = n \cdot m$$

which shows that $\oplus$ and $+$ coincide when applied to natural numbers, and similarly $\circ$ and $\cdot$. This is some justification for speaking, as I did earlier, of *adding* and *multiplying* fractions in much the same sense as we speak of adding and multiplying natural numbers. It also encourages us to pursue the analogy between fractions and natural numbers further, and perhaps the next thing to consider is whether we can introduce a relation between fractions corresponding to $>$ between natural numbers.

One thing that seems fairly clear is that we shall want to ensure that if

$$\frac{n}{m} > \frac{p}{q}$$

then whatever is n m^{ths} of something must be greater than whatever is p q^{ths} of it, i.e. (for each subscripted ϕ)

$$(\forall xyz)(x \approx \frac{n}{m}{\circ}z \ \& \ y \approx \frac{p}{q}{\circ}z \rightarrow x \succ y).$$

Conversely, if it holds (for even one subscripted ϕ) that

$$(\exists xyz)(x \approx \frac{n}{m}{\circ}z \ \& \ y \approx \frac{p}{q}{\circ}z \ \& \ x \succ y)$$

then we shall expect it to follow that

$$\frac{n}{m} > \frac{p}{q}$$

for both these laws are easily established for $>$ between natural numbers. Now on investigation it turns out that these suggested necessary and sufficient conditions are equivalent to each other and to the simple condition

$$n \cdot q > p \cdot m$$

but the proof of this is not entirely elementary. In fact we must first establish an important theorem on divisibility, namely for *every* quantity ϕ-ness, either *every* ϕ is divisible as far as we please, or *every* ϕ is an integral multiple of some 'unit' ϕ. That is, for each subscripted ϕ.

$$(\forall x)(\forall n)(\exists y)(x \approx n{\circ}y) \lor (\exists y)(\forall x)(\exists n)(x \approx n{\circ}y).$$

In the first case I shall say that ϕ-ness is a *continuous* quantity[55], and in the second that it is *discrete*. (The disjunction is evidently exclusive.) Since this theorem is of fundamental importance, it is worth while pausing here to sketch a proof of it.

The proof breaks into three stages, and we establish in order:

(A) $(\exists y)(\forall x)(\exists n)(x \approx n{\circ}y) \lor (\forall x)(\exists y)(x \succ y)$

(B) $(\forall x)(\exists y)(x \succ y) \rightarrow (\forall x)(\forall n)(\exists y)(x \succ n{\circ}y)$

(C) $(\forall x)(\forall n)(\exists y)(x \succ n{\circ}y) \rightarrow (\forall x)(\forall n)(\exists y)(x \approx n{\circ}y).$

Both (A) and (B) are straightforward applications of Archimedes' postulate. For example to establish (A) we assume first that some element a is a *minimal* element, i.e. that

(i) $(\forall x)(x \succcurlyeq a).$

Now suppose also (for *reductio ad absurdum*) that an element b is not a multiple of this minimal element, i.e. that

(ii) $(\forall n)(b \not\approx n{\circ}a).$

Taking $n = 1$, (i) and (ii) together obviously imply

(iii) $b \succ 1{\circ}a$

---

[55] It will be shown later that if ϕ-ness is a continuous quantity then every ϕ can be divided in any desired proportion, rational or irrational (pp. 237–9, 248–53).

which we may use as a basis for an inductive argument. For now if we assume as inductive hypothesis

(iv) $b \succ n{\circ}a$

it follows that for some c

$$b \approx n{\circ}a \oplus c$$

However, by (i) we have $c \succcurlyeq a$, and hence

$$b \succcurlyeq n{\circ}a \oplus a$$

that is

$$b \succcurlyeq n'{\circ}a.$$

Applying (ii) to this result we infer

(v) $b \succ n'{\circ}a$

which is the desired conclusion. For from (iii) and the fact that (iv) implies (v) we infer

$$(\forall n)(b \succ n{\circ}a)$$

in direct contradiction to Archimedes' postulate. Therefore (i) and (ii) do indeed yield a contradiction and we have shown that

$$(\forall x)(x \succcurlyeq a) \rightarrow (\exists n)(b \approx n{\circ}a)$$

from which (A) above follows at once. The argument for (B) is very similar, so let us now turn our attention to (C). We begin by assuming

(i) $(\forall x)(\forall n)(\exists y)(x \succ n{\circ}y).$

From this we must now establish two lemmas:

(ii) $a \succ n{\circ}b \rightarrow (\exists x)(a \succ n{\circ}x \ \& \ x \succ b)$

(iii) $a \not\succcurlyeq n{\circ}b \rightarrow (\exists x)(a \not\succcurlyeq n{\circ}x \ \& \ x \not\succcurlyeq b).$

The first is very simply obtained, for if

$$a \succ n{\circ}b$$

then there is some element, say c, such that

$$a \approx n{\circ}b \oplus c.$$

However, according to (i)

$$(\exists y)(c \succ n \circ y)$$

and therefore

$$(\exists y)(a \succ (n \circ b) \oplus (n \circ y))$$

$$(\exists y)(a \succ n \circ (b \oplus y) \ \& \ b \oplus y \succ b)$$

$$(\exists x)(a \succ n \circ x \ \& \ x \succ b).$$

So lemma (ii) presents no difficulty. By an entirely similar argument we could easily establish

$$a \prec n \circ b \to (\exists x)(a \prec n \circ x \ \& \ x \prec b)$$

but this is a weaker proposition than lemma (iii) as stated, for in this proposition the antecedent implies E!$n \circ b$, while in lemma (iii) this is not so. Consequently the argument for (iii) is somewhat more roundabout. Contraposing (iii) as stated, to eliminate the negated inequalities, what has to be proved is

$$(\forall x)(b \succ x \to a \succcurlyeq n \circ x) \to a \succcurlyeq n \circ b.$$

It is convenient to distinguish the cases $n = 1$ and $n \neq 1$, and I take the second case first, where the antecedent can be written as

$$(\forall x)(b \succ x \to a \succcurlyeq n' \circ x).$$

Now assume

$$b \succ c.$$

The difference $b \ominus c$ therefore exists, and using our premise (i) we may infer that for some x

$$b \ominus c \succ n' \circ x$$

which is to say

$$b \succ n \circ x \oplus x \oplus c$$

and therefore

$$b \ominus x \succ n \circ x \oplus c.$$

Clearly we have $b \succ b \ominus x$, so we can use the antecedent to deduce

$$a \succcurlyeq n' \circ (b \ominus x)$$

and putting this together with the last we have

$$a \succ n \circ (b \ominus x) \oplus n \circ x \oplus c$$

or in other words

$$a \succ n{\circ}b \oplus c.$$

Conditionalizing and generalizing the argument so far, we have therefore shown that our antecedent implies

$$(\forall x)(b \succ x \rightarrow a \succ n{\circ}b \oplus x).$$

From here it is quite easy to complete the argument. By our premise (i) we have $(\exists y)(b \succ y)$, and therefore $(\exists y)(a \succ n{\circ}b \oplus y)$, and therefore

$$(\exists x)(a \approx n{\circ}b \oplus x).$$

Applying our last result once more, this yields

$$(\exists x)(a \approx n{\circ}b \oplus x \ \& \ x \succcurlyeq b)$$

and hence

$$a \succcurlyeq n{\circ}b \oplus b$$

which is the desired conclusion

$$a \succcurlyeq n'{\circ}b.$$

This establishes our lemma (iii) for the case $n \neq 1$. In case $n = 1$ the above argument may be repeated by writing 1 for n' throughout and simply deleting terms beginning with n.

The completes the proof of the two lemmas, and we may now proceed to the proof of our main theorem (C), which is achieved by an appeal to the principle of least upper bounds. Clearly, for each n, the elements x such that $a \succ n{\circ}x$ will have an upper bound (namely a) and so by L.U.B. they will have a least upper bound, say b, such that

(iv) $(\forall z)(b \preccurlyeq z \leftrightarrow (\forall x)(a \succ n{\circ}x \rightarrow x \preccurlyeq z)).$

Using the implication from left to right we have (since $b \preccurlyeq b$)

$$(\forall x)(a \succ n{\circ}x \rightarrow x \preccurlyeq b).$$

$\therefore \quad \sim(\exists x)(a \succ n{\circ}x \ \& \ x \succ b)$

and therefore, by (ii),

(v) $a \not\succ n{\circ}b.$

But also, using the implication from right to left, we have

$$(\forall z)(b \succ z \rightarrow (\exists x)(a \succ n{\circ}x \ \& \ x \succ z))$$

$\therefore \quad (\forall z)(b \succ z \rightarrow a \succ n{\circ}z)$

$\therefore \quad (\forall z)(z \succcurlyeq b \rightarrow a \succcurlyeq n{\circ}z)$

$\therefore \quad {\sim}(\exists z)(z \succcurlyeq b \ \& \ a \succcurlyeq n{\circ}z)$

and therefore, by (iii)

(vi) $a \succcurlyeq n{\circ}b$.

However, (v) and (vi) together evidently yield

$$a \approx n{\circ}b$$

and so, discharging assumption (iv), we deduce

(vii) $(\exists y)(a \approx n{\circ}y)$

which is the desired conclusion since a was any element and n any number. This completes the proof.

With this theorem on divisibility to hand we may return to our investigation of the 'abstraction'

$$\frac{n}{m} > \frac{p}{q}$$

and its two suggested conditions

(I) $(\exists xyz)(x \approx \dfrac{n}{m}{\circ}z \ \& \ y \approx \dfrac{p}{q}{\circ}z \ \& \ x \succ y)$

(II) $(\forall xyz)(x \approx \dfrac{n}{m}{\circ}z \ \& \ y \approx \dfrac{p}{q}{\circ}z \rightarrow x \succ y)$.

Let us begin by showing that the first of them implies

(III) $n{\cdot}q > p{\cdot}m$.

We assume, then, that (I) holds for some quantity ϕ-ness. Then from the definition of $(n/m){\circ}z$ it is easily shown that there must be objects a, b (namely $(1/m){\circ}z$ and $(1/q){\circ}z$ respectively) such that

(i) $n{\circ}a \succ p{\circ}b \ \& \ m{\circ}a \approx q{\circ}b$.

Now we consider two cases according as ϕ-ness is a discrete or a continuous quantity. In the first case suppose ϕ-ness is discrete, so that a and b are each integral multiples of some 'unit' ϕ, say c. That is

(ii) $(\exists kl)(a \approx k \circ c \ \& \ b \approx l \circ c)$.

Putting (i) and (ii) together we have

$$(\exists kl)(n \circ (k \circ c) \succ p \circ (l \circ c) \ \& \ m \circ (k \circ c) \approx q \circ (l \circ c))$$

$\therefore \quad (\exists kl)((n \cdot k) \circ c \succ (p \cdot l) \circ c \ \& \ (m \cdot k) \circ c \approx (q \cdot l) \circ c)$

$\therefore \quad (\exists kl)(n \cdot k > p \cdot l \ \& \ m \cdot k = q \cdot l)$

$\therefore \quad (\exists kl)(n \cdot k \cdot q > p \cdot l \cdot q \ \& \ p \cdot m \cdot k = p \cdot q \cdot l)$

$\therefore \quad (\exists k)(n \cdot k \cdot q > p \cdot m \cdot k)$

$\therefore \quad n \cdot q > p \cdot m$

which is the desired result. In the second case suppose ϕ-ness is continuous, so that a may be divided into q^{ths} and b into m^{ths}. That is

(iii) $(\exists xy)(a \approx q \circ x \ \& \ b \approx m \circ y)$.

Putting (i) and (iii) together we have

$$(\exists xy)(n \circ (q \circ x) \succ p \circ (m \circ y) \ \& \ m \circ (q \circ x) \approx q \circ (m \circ y))$$

$\therefore \quad (\exists xy)((n \cdot q) \circ x \succ (p \cdot m) \circ y \ \& \ (m \cdot q) \circ x \approx (m \cdot q) \circ y)$

$\therefore \quad (\exists xy)((n \cdot q) \circ x \succ (p \cdot m) \circ y \ \& \ x \approx y)$

$\therefore \quad (\exists x)((n \cdot q) \circ x \succ (p \cdot m) \circ x)$

$\therefore \quad n \cdot q > p \cdot m.$

Therefore the desired conclusion follows from either hypothesis, and we infer that if (I) holds even for one subscripted ϕ then (III) holds also. By an entirely similar argument it may be shown that if (III) holds then (II) must hold for every subscripted ϕ. Thus (I) implies (III) and (III) implies (II).

To complete the argument we have to show that (II) implies (I), and this evidently depends upon our being able to show that at least for some subscripted ϕ

$$(\exists xyz)(x \approx \frac{n}{m} \circ z \ \& \ y \approx \frac{p}{q} \circ z).$$

But this is seen to be quite obvious when we recognize that the fractions we are talking of cover, *inter alia*, fractions *of natural numbers themselves.* Admittedly I started by discussing fractions of such things as apples, rather than fractions of numbers, and it may well seem that a somewhat different sense of 'fraction' is involved here. Indeed I would myself agree, and I would contend that the former is the primary and most usual sense of 'fraction', and this sense requires the notion of a part for its analysis. However, the course we actually took deliberately left that aspect of fractions out of account. Our formal definitions were couched solely in terms of $\geqslant$ and n°, and therefore they apply equally well to any system of entities that satisfies the traditional axioms in terms of $\oplus$. Hence they apply perfectly well to the natural numbers themselves, where of course $\approx$ and $>$ are taken as $=$ and $>$ between numbers, $\oplus$ is taken as $+$ between numbers, and n° is taken as $n\cdot$. But it is obvious that

$$n\cdot q = \frac{n}{m}\cdot(m\cdot q) \ \& \ p\cdot m = \frac{p}{q}\cdot(m\cdot q)$$

$$\therefore \qquad (\exists l_1 l_2 k)(l_1 = \frac{n}{m}\cdot k \ \& \ l_2 = \frac{p}{q}\cdot k).$$

Therefore our existential thesis certainly holds for 'this subscripted ϕ'. This completes the proof[56].

So far, then, we have suggested obtaining three sorts of formulae which have the appearance of singular statements about fractions, namely

$$\frac{n}{m} = \frac{p}{q} \oplus \frac{l}{k}$$

---

[56] If we wish to revert to the original approach, in which the notion of a part was taken as fundamental, then we may take $\approx$ as 'as many as' between *finite* (non-empty) sets of natural numbers, and inclusion as our part–whole relation. This interpretation certainly satisfies the original axioms P1, P2, C1, C2 and (ADD +)∗. Then we may easily show that

$$\{k:k\leqslant n\cdot q\} \approx \frac{n}{m}\circ\{k:k\leqslant m\cdot q\} \ \& \ \{k:k\leqslant p\cdot m\} \approx \frac{p}{q}\circ\{k:k\leqslant m\cdot q\}.$$

The set-theoretical notation can be construed syncategorematically (i.e. in accordance with Russell's no-class theory) throughout.

$$\frac{n}{m} = \frac{p}{q} \circ \frac{l}{k}$$

$$\frac{n}{m} > \frac{p}{q}$$

as abbreviations for generalizations in which those fractions are applied, respectively

$$(\forall xy)(x \approx \frac{n}{m} \circ y \leftrightarrow x \approx \frac{p}{q} \circ y \oplus \frac{l}{k} \circ y)$$

$$(\forall xy)(x \approx \frac{n}{m} \circ y \leftrightarrow x \approx \frac{p}{q} \circ \frac{l}{k} \circ y)$$

$$(\forall xyz)(x \approx \frac{n}{m} \circ z \,\&\, y \approx \frac{p}{q} \circ z \rightarrow x \succ y)$$

(where, in each case, the omitted subscript ϕ is also understood to be universally generalized). In the last case our formula behaves just as the familiar formula of rational arithmetic, since (as we have just seen)

$$\frac{n}{m} > \frac{p}{q} \leftrightarrow n \cdot q > p \cdot m.$$

However, in the first two cases the correspondence is more limited and holds (roughly speaking) only when the rational numbers in question are expressed 'in their lowest terms'. It is easy enough to improve the correspondence if we wish to, for on our present approach there are evidently fractions, e.g. one half (of) and two-quarters (of), which are not the *same* fractions but which are also such that neither is greater than the other—a situation that does not occur with natural numbers. We may well be tempted to call such fractions *equal*, and it would be quite a natural step to define, by analogy with our definition for $>$ between fractions,

$$\frac{n}{m} \approx \frac{p}{q}$$

for

$$(\forall xyz)(x \approx \frac{n}{m} \circ z \,\&\, y \approx \frac{p}{q} \circ z \rightarrow x \approx y).$$

It will then emerge as a consequence that

$$\frac{n}{m} \approx \frac{p}{q} \leftrightarrow n \cdot q = p \cdot m.$$

Hence it is easy to see that if we write our 'arithmetic of fractions' by using $\approx$ everywhere instead of $=$ the resulting system will be indistinguishable from the usual system of rational arithmetic.

But why should we want to do this? Why obliterate the apparently useful distinction between a half and two quarters? For even though we may still call these *equal* fractions the fact will remain that they are not the *same* fraction in so far as we cannot always substitute one for the other in any application. I suspect that the pressure comes from a different use of rational numbers, which we may begin by describing as their use to characterize *ratios* rather than *fractions*. If a is $\frac{1}{2}$ of b, we have said, it does not follow that a is also $\frac{2}{4}$ of b, since b might not have quarters. On the other hand if the ratio of a to b (e.g. in size) is that of 1 to 2 it surely does follow that it is also that of 2 to 4, since the ratios of 1 to 2 and of 2 to 4 *are the same ratio*. So let us now move on to consider ratios.

Perhaps the first question that might be raised about ratios is this: *Why* do we say that, for example, the ratio of 2 to 1 and the ratio of 4 to 2 are the *same* ratio? What is the same about them? An orthodox answer might perhaps be that these are the same ratio just because $2 \cdot 2 = 4 \cdot 1$, for generally the ratio of n to m is the same as the ratio of p to q if and only if $n \cdot q = p \cdot m$. However, I find that on reflection that this answer is not satisfying. Why *should* this be the criterion for sameness of ratio? How does it show that there is anything the *same* about these ratios? Of course one must admit that the relation which holds between two ratios, n to m and p to q, if and only if $n \cdot q = p \cdot m$ is certainly an equivalence relation. So one might with some reason claim that it must pick out *some* respect in which these ratios are the same. But that is clearly not enough to show that this particular sameness is actually sameness of ratio. Let us see whether we can get any deeper.

If one asks why the ratio of 2 to 1 is the same as the ratio of 4 to 2, I think that it is more satisfying to be told that this is because 2 *is twice* 1 and 4 *is twice* 2, or—putting it in a more

long winded manner—2 is the same multiple of 1 as 4 is of 2.
Similarly it seems quite satisfying to explain that, for example,
the ratio of 3 to 6 is the same as that of 4 to 8 just because 3
is half of 6 and 4 is half of 8, which is to say that each is the
same fraction of the other. If we permit improper fractions, as
indeed we have been doing, then multiples may be counted a
special case of fractions, and so we may say quite generally that
the ratio of n to m is the same as that of p to q if and only if
n is the same fraction of m as p is of q. What is important is
clearly that n and m stand in the same relation of magnitude
(i.e. in the same relations of numerical comparison) as do p and
q, but so long as we are confining our attention to ratios between
natural numbers these relations of magnitude *can* always be expressed by fractions, just because every natural number is some
definite fraction of every other. Using the notation '$n:m = p:q$'
to express 'the ratio of n to m is the same as that of p to q'—or,
more briefly, 'n is to m as p is to q'—we may therefore define

$$n:m = p:q \quad \text{for} \quad (\exists kl)(n = \frac{k}{l} \cdot m \ \& \ p = \frac{k}{l} \cdot q).$$

From this definition we can easily recover the orthodox
criterion by proving that

$$(\exists kl)(n = \frac{k}{l} \cdot m \ \& \ p = \frac{k}{l} \cdot q) \leftrightarrow n \cdot q = p \cdot m.$$

The argument from left to right is very straightforward, for
since[57]

$$n = \frac{k}{l} \cdot m \rightarrow l \cdot n = k \cdot m$$

the left-hand side at once yields

$$(\exists kl)(l \cdot n = k \cdot m \ \& \ l \cdot p = k \cdot q)$$

$$\therefore \qquad (\exists kl)(l \cdot n \cdot q = k \cdot m \cdot q \ \& \ l \cdot p \cdot m = k \cdot q \cdot m)$$

---

[57] In general

$$a \approx \frac{n}{m} \circ b \rightarrow m \circ a \approx n \circ b$$

requires the hypothesis $E\,!\,m \circ a \lor E\,!\,n \circ b$, but of course the hypothesis is satisfied
in this case.

$\therefore \qquad (\exists l)(l \cdot n \cdot q = l \cdot p \cdot m)$

$\therefore \qquad n \cdot q = p \cdot m.$

As for the converse, we rely here upon the lemma cited on p. 170, which states, when applied to natural numbers, that

$$Prime(l,\, k) \rightarrow (l \cdot n = k \cdot m \rightarrow n = \frac{k}{l} \cdot m).$$

For now if we assume the right-hand side of our equivalence, viz

(i) $n \cdot q = p \cdot m$

we know that p and q will certainly have a highest common factor, say i, such that for some $k,\, l$

(ii) $p = k \cdot i$ & $q = l \cdot i$ & $Prime(l,\, k)$.

From (ii) we have at once

$$l \cdot p = k \cdot q$$

and from (i) and (ii) together we have

$$l \cdot n = k \cdot m$$

and so, since we *are* now given $Prime(l,\, k)$, we infer from these

$$n = \frac{k}{l} \cdot m \;\&\; p = \frac{k}{l} \cdot q$$

which was to be proved. In this way we recover the orthodox criterion

$$n : m = p : q \leftrightarrow n \cdot q = p \cdot m.$$

But the detour has been worthwhile if it sheds any light on *why* this criterion should pick out sameness of ratio.

Next we may extend our reasoning to cover inequalities of ratios. Applying the same reasoning as before, the ratio of n to m will be greater than that of p to q if and only if n is a greater fraction of m than p is of q, so we may define

$$n : m > p : q \quad \text{for} \quad (\exists k_1 k_2 l_1 l_2)(n = \frac{k_1}{l_1} \cdot m \;\&\; p = \frac{k_2}{l_2} \cdot q \;\&\; \frac{k_1}{l_1} > \frac{k_2}{l_2}).$$

Using our previous result on when one fraction is greater than another we therefore infer

$$n:m > p:q \leftrightarrow (\exists k_1 k_2 l_1 l_2)(n = \frac{k_1}{l_1} \cdot m \ \& \ p = \frac{k_2}{l_2} \cdot q \ \& \ k_1 \cdot l_2 > l_1 \cdot k_2)$$

and hence, by an argument similar to the preceding,

$$n:m > p:q \leftrightarrow n \cdot q > p \cdot m$$

which is again the orthodox criterion. We thus have an entirely natural analysis of statements comparing ratios between natural numbers, and on this analysis it turns out that while greater ratios correspond exactly to greater fractions, identical ratios correspond not to identical fractions but to equal fractions. So we have certainly approached more closely to the usual views about rational numbers.

To fill out this 'arithmetic of ratios' we should now go on to consider the addition and multiplication of ratios, but at least at first glance the whole idea of adding and multiplying ratios may seem rather strange. Why should we suppose that two ratios can somehow be *added* to form a further ratio? To answer this question we shall I think have to turn back again to the *application* of rational numbers to things other than numbers, and indeed to leave the idea of a ratio on one side for a while. As a first step, however, let us ask how ratios between things other than numbers are compared with ratios between numbers, as when it is said that the ratio of a to b in ϕ-ness is (or is greater than or less than) that of n to m. I shall abbreviate this to $a :_\phi b = n:m$, and similarly for $>$ and $<$, and I continue to omit the subscript ϕ for brevity.

At the outset we are met by a difficulty, which is that our account of ratios between natural numbers relied upon the fact that each natural number is a definite fraction of every other, which is to say that the natural numbers are all *commensurable*. Now as we saw from our theorem on divisibility, the instances of any *discrete* quantity must again be all commensurable (in respect of that quantity), but with *continuous* quantities this is not so. In fact this point does not disturb our account of *sameness* of ratio, for of course if the ratio of a to b is that of n to m then a and b must certainly be commensurable. So in this case we may simply generalize our previous definition by putting

$$a:b = n:m \quad \text{for} \quad (\exists pq)(a \approx \frac{p}{q} \circ b \ \& \ n = \frac{p}{q} \cdot m).$$

The difficulty emerges when we come to inequalities of ratios. For a discrete quantity it will certainly hold as before that

$$a:b \geqslant n:m \leftrightarrow (\exists pqkl)(a \approx \frac{p}{q} \circ b \ \& \ n = \frac{k}{l} \cdot m \ \& \ \frac{p}{q} \gtrsim \frac{k}{l})$$

which we might more briefly express as[58]

(A) $a:b \geqslant n:m \leftrightarrow (\exists pq)(\frac{1}{p} \circ a \approx \frac{1}{q} \circ b \ \& \ p:q \geqslant n:m).$

But this evidently will not hold where a and b are incommensurable, as they may well be in the continuous case. On the other hand in the continuous case a and b may always be divided into as minute fractions as we please, so in this case we have very simply

(B) $a:b \geqslant n:m \leftrightarrow \frac{1}{n} \circ a \geqslant \frac{1}{m} \circ b.$

However, this does not hold in the discrete case. The way out of this dilemma that has been orthodox since the first treatment of the question by Eudoxus[59] has been to employ a variant of equivalence (B) but with *multiples* of a and b rather than *fractions*, defining

$$a:b \geqslant n:m \quad \text{for} \quad m \circ a \geqslant n \circ b.$$

However, this course is not open to us, for just as we do not assume that every quantity is continuous (so that fractions always exist, however small) so we do not assume that every quantity is unbounded (so that multiples always exist, however large), and the orthodox definition clearly gives the wrong

---

[58] Notice, incidentally, that the definition of sameness of ratio implies

$$a:b = n:m \leftrightarrow (\exists pq)(\frac{1}{p} \circ a \approx \frac{1}{q} \circ b \ \& \ p:q = n:m).$$

[59] Eudoxus' treatment is preserved for us in Euclid, *Elements*, Book V (for the attribution to Eudoxus see, for example, Heath's commentary, *ad loc.*).

results if the multiples in question fail to exist. We must there-fore compromise between (A) and (B), and the most obvious suggestion is[60]

$$a:b \geqslant n:m \quad \text{for} \quad (\exists pq)(\frac{1}{p} \circ a \succcurlyeq \frac{1}{q} \circ b \ \& \ p:q \geqslant n:m)$$

to which we may add, dually,

$$a:b \leqslant n:m \quad \text{for} \quad (\exists pq)(\frac{1}{p} \circ a \preccurlyeq \frac{1}{q} \circ b \ \& \ p:q \leqslant n:m).$$

The adequacy of these definitions is confirmed by noticing that they yield the expected ordering properties, for example

$$a:b \geqslant n:m \lor a:b \leqslant n:m$$

$$a:b \geqslant n:m \ \& \ a:b \leqslant n:m \rightarrow a:b = n:m$$

$$a:b \geqslant n:m \ \& \ a:b \leqslant p:q \rightarrow p:q \geqslant n:m$$

which are all quite simply established both for discrete and for continuous quantities. We may of course add the definitions

$$a:b > n:m \quad \text{for} \quad a:b \geqslant n:m \ \& \ a:b \nleqslant n:m$$

$$a:b < n:m \quad \text{for} \quad a:b \leqslant n:m \ \& \ a:b \ngeqslant n:m$$

and no further difficulties arise in thus 'applying' ratios between natural numbers.

We are now in a position to return to our original subject, relations of numerical comparison, for it seems to me that a statement of numerical comparison of the form '*a* is (at least) *n/m* times as φ as *b*' is exactly equivalent to the explicit ratio-statement 'the ratio of *a* to *b* in φ-ness is (at least as great as) that of *n* to *m*'. Clearly the comparisons cannot be construed as remarks directly about fractions. For example '*a* is at least $\frac{3}{4}$ times as numerous as *b*' will be true when (say) *a* is a class of

[60] More long windedly, but more in keeping with our treatment of the numerical case, we could equivalently write the first definiens here as

$$(\exists pqkl)(a \succcurlyeq \frac{p}{q} \circ b \ \& \ \frac{p}{q} \succsim \frac{k}{l} \ \& \ n = \frac{k}{l} \cdot m).$$

A similar rewriting is *not* available for the second (but see footnote 61). Technically, the more long winded definition is clearly preferable, for other-wise the formula $p:q \geqslant n:m$, which is a special case of the definiendum, must here be being given a second and circular definition in terms of itself.

7 members and b a class of 9 members, though there is nothing here that is a quarter of b or a third of a. In this context all that is important is the *ratio* of 3 to 4, and we could perfectly well say instead 'the ratio of a to b in "numerosity" is at least as great as that of 3 to 4', i.e. '$a:b \geqslant 3:4$'. Let us take it, then, that we may move freely between statements explicitly about ratios and statements of numerical comparison. In that case it will be natural to introduce a new way of expressing our ratio-statements, for to correspond to our notation '$a \succcurlyeq_n b$' for 'a is at least n times as ϕ as b' we shall naturally employ '$a \succcurlyeq_{(n/m)} b$' to express 'a is at least n/m times as ϕ as b', and the relevant definitions are simply[61]

$$a \succcurlyeq_{(n/m)} b \quad \text{for} \quad a:b \geqslant n:m$$

$$a \preccurlyeq_{(n/m)} b \quad \text{for} \quad a:b \leqslant n:m.$$

Statements of numerical comparison are then to be distinguished from statements directly involving fractions (or multiples; recall that $a \preccurlyeq_n b$ does not imply $E!n{\circ}b$ just as $a \succcurlyeq_{(n/m)} b$ does not imply $E!(n/m){\circ}b$), but the two are alike in strongly suggesting definitions of addition and multiplication. Thus where only natural numbers were involved we could easily show that

$$a \approx_n c \;\&\; b \approx_m c \rightarrow a \oplus b \approx_{n+m} c$$

$$a \approx_n c \;\&\; c \approx_m b \rightarrow a \approx_{n \cdot m} b$$

(provided, in the first case, $E!a \oplus b$). Now that rational numbers are permitted in the same contexts, we can quite easily show from the results previously noted about fractions that (with the same proviso)

$$a \approx_{(n/m)} c \;\&\; b \approx_{(p/q)} c \rightarrow a \oplus b \approx_{(n \cdot q + p \cdot m/m \cdot q)} c$$

$$a \approx_{(n/m)} c \;\&\; c \approx_{(p/q)} b \rightarrow a \approx_{(n \cdot p/m \cdot q)} b.$$

This is evidently a powerful motive for adopting the orthodox definitions

$$\frac{n}{m} + \frac{p}{q} \quad \text{for} \quad \frac{n \cdot q + p \cdot m}{m \cdot q}$$

---

[61] Alternatively we could define $\preccurlyeq_{(n/m)}$ in terms of $\succcurlyeq_{(n/m)}$, much as $\preccurlyeq_n$ was defined in terms of $\succcurlyeq_n$.

$$\frac{n}{m} \cdot \frac{p}{q} \quad \text{for} \quad \frac{n \cdot p}{m \cdot q} \ .$$

We saw in §1 that the classical accounts of rational arithmetic can justify these definitions only by a wish to preserve for rational numbers several of the laws that hold in the arithmetic of natural numbers (and in particular the law of distribution). However, I think an equally strong, and indeed stronger, justification is that the definitions preserve for the *application* of rational numbers laws that hold in the corresponding *application* of natural numbers.

Supposing, then, that these definitions are adopted for statements of numerical comparison, there is evidently a motive for transferring them also to explicit ratio-statements in view of the general interchangeability of the two sorts of statement. So we may well define

$$(n:m) + (p:q) \quad \text{for} \quad (n \cdot q + p \cdot m):(m \cdot q)$$

$$(n:m) \cdot (p:q) \quad \text{for} \quad (n \cdot p):(m \cdot q)$$

just in order to keep this interchangeability. This would complete the foundation of our 'arithmetic of ratios' to bring it more into line with the usual arithmetic of rational numbers.

However, it seems to me that neither this 'arithmetic of ratios' nor the earlier 'arithmetic of fractions' quite corresponds to what we ordinarily think of as the arithmetic of rational numbers. In the case of fractions the clearest difference was that there is good reason to distinguish between identity and equality of fractions, but this distinction is not generally held to apply to rational numbers. In the case of ratios there is still, I think, some force in the contention that we do not find it as natural to add and multiply ratios as we do to add and multiply rational numbers, but a clearer difference is this: ratios between natural numbers are surely quite different in kind from the natural numbers themselves, and formulae such as $4:2 = 2$ or $3:2 > 1$ seem to make no sense at all. On the other hand formulae such as $4/2 = 2$ or $3/2 > 1$ do not occasion any surprise as formulae of rational arithmetic, for we simply do think of the rational numbers as falling into an order which is quite definite with respect to the natural numbers; the natural numbers are themselves part of this ordering.

What this suggests is that the arithmetic of rational numbers is much better explained as being obtained by abstraction from generalizations concerning numerical comparisons in just the way in which we suggested that an arithmetic of fractions could be obtained by abstraction from the corresponding generalizations applying to fractions. More simply than in the case of fractions we might adopt as our basic definition that

$$\frac{n}{m} \geqslant \frac{p}{q}$$

is to abbreviate the generalization that for all subscripted ϕ

$$(\forall xy)(x \geqslant_{(n/m)} y \to x \geqslant_{(p/q)} y)$$

and it will then be entirely natural to extend the same abbreviation to cases where one or both of n/m and p/q are supplanted by expressions for natural numbers. Ordinary elementary rational arithmetic may then be developed in a perfectly obvious way. In view of our previous results, the definition is easily seen to yield the consequence

$$\frac{n}{m} \geqslant \frac{p}{q} \leftrightarrow n \cdot q \geqslant m \cdot p$$

Addition and multiplication of rational numbers is already defined by the use of $+$ and $\cdot$ in numerical comparisons, and other relations between rational numbers (and between rational and natural numbers) are evidently definable in terms of $\geqslant$. If we wish to use variables, say r, s, . . ., ranging directly over rational numbers then these can be explained as on p. 79 by defining

$$(\forall r)(-r-)$$

as short for

$$(\forall nm)(-\frac{n}{m}-).$$

No difficulties are encountered. We can for example prove all the unargued postulates used on pp. 82–4 in developing the formalist account. In sum, if rational numbers are construed as being obtained by abstraction from numerical comparisons, so that the statements of rational arithmetic are construed as generalizations over numerical comparisons, then this construal actually dictates all fundamental definitions or axioms of rational arithmetic.

I postpone further comment on the philosophical interest of this construction until Chapter 4. Meanwhile I bring the present chapter to a close with a postscript on one or two technical details that have been rather lightly passed over. Our fundamental definition has been given, in a somewhat cavalier manner, by saying that

$$\frac{n}{m} \geqslant \frac{p}{q}$$

is to abbreviate the statement that for all subscripted ϕ

$$(\forall xy)(x \succcurlyeq_{(n/m)} y \to x \succcurlyeq_{(p/q)} y).$$

A minor point may be made first, namely that from the formal point of view it will be necessary to tag the sign $\geqslant$ hereby introduced, say as $\geqslant_r$, in order to distinguish it from the sign $\geqslant$ already in use between expressions for natural numbers (and perhaps from a corresponding sign used between expressions for fractions). For when the definition above is extended to cover natural numbers as well as rational numbers it will become a significant (but easily proved) thesis that

$$n \geqslant_r m \leftrightarrow n \geqslant m.$$

(The same considerations obviously apply to $=_r$, $>_r$, $+_r$, $\cdot_r$, and so on).

More importantly, we should say something in elaboration of the rather cryptic phrase 'for all subscripted ϕ'. Clearly what is intended here is that whenever we have a relation $\succcurlyeq_\varphi$ satisfying the axioms for comparative relations C1 and C2, and whenever we have a binary quasi-operation $\oplus$ and a set of unary quasi-operations n° satisfying the traditional axioms for extensive quantities, and whenever $\succcurlyeq_{(n/m)}$ and $\succcurlyeq_{(p/q)}$ are defined in terms of these according to the definitions given, then … (the displayed formula holds). To fill in the details explicitly it is simpler to take the traditional postulates as formulated for triadic relations, as on p. 156. Let us now use ϕ directly as a variable for triadic relations (superseding the earlier conventions). It is to be a type-neutral variable, ranging over homogeneous triadic relations of any type, for it is this which legitimizes the special case 'of the subscripted ϕ' in which the variables are taken as natural number variables. For convenience we may continue to use $\phi\alpha$ as a short way of saying that α is in the field of ϕ, abbreviating

$$\phi\alpha \quad \text{for} \quad (\exists\beta\gamma)(\phi(\alpha, \beta, \gamma) \vee \phi(\beta, \alpha, \gamma) \vee \phi(\beta, \gamma, \alpha)).$$

We then rewrite (the closures of) postulates (a)–(e) so that the variables are restricted everywhere to entities α such that $\phi\alpha$ (compare footnote 40), and abbreviate the resulting conjunction to

$$Sat(\phi).$$

It is also convenient to abbreviate further

$$Sat(\alpha, \beta, \phi) \quad \text{for} \quad \phi\alpha \ \& \ \phi\beta \ \& \ Sat(\phi).$$

The subscript ϕ is added everywhere in our series of definitions, and the definition of

$$\frac{n}{m} \geqslant_r \frac{p}{q}$$

is then given as

$$(\forall\phi)(\forall\alpha\beta)(Sat(\alpha, \beta, \phi) \to (\alpha \succcurlyeq_{(n/m)\varphi} \beta \to a \succcurlyeq_{(p/q)\varphi} \beta)).$$

If we were to construe the whole string of definitions as nothing more than abbreviations, then of course we should find that the theses of rational arithmetic appear now as purely logical truths conveniently abbreviated[62]. However, as I maintained in Volume I (Chapter 6, §2) that is actually a quite misleading way of construing them. The route that we have taken is intended as a piece of *analysis*, and an analysis is a great deal more than a string of abbreviations. The analysis in fact stretches back to §3 of this chapter, and does not begin with an unargued set of postulates in §6.

---

[62] Note that the principle of dependent choices has not been used since L.U.B. was adopted as an axiom.

3. Real Numbers

1. Classical Accounts: Cuts and Sequences[1]

I began my discussion of 'classical' accounts of rational numbers by pointing out that every rational number is determined by an ordered pair of natural numbers, and conversely that every ordered pair of natural numbers determines a rational number. It is not perhaps so obvious how real numbers may be thouhgt of as determined by rational numbers, and indeed there seems to be no *one* method of determination which suggests itself as the most natural.

Perhaps the simplest approach (from the point of view of more or less schoolboy arithmetic) is obtained by using the decimal notation. When rational numbers are expressed in decimal notation it is soon discovered that while in some cases the decimal expansion comes to an end, in others it goes on for ever. In the latter case the expansion will in fact always be periodic, i.e. after some point in the expansion there will be a digit or group of digits that recurs without interruption, so that the expansion falls into a regular and predictable pattern. This is not, however, immediately obvious, and we are generally quite happy to operate with the idea of an unending decimal before realizing this. So the real numbers can naturally be regarded as comprising all those numbers that can be written in the decimal notation (granting the fiction that an unending decimal can in fact be 'written'), and we then learn—perhaps to our surprise—that not all such numbers are rational. This way of approaching the real numbers doubtless is the way that many

---

[1] Since the classical accounts (e.g. Cantor 1883, Dedekind 1872) are framed in terms of *signed* real numbers (and signed rational numbers), I too shall be speaking of signed numbers in the present section, though elsewhere in the book my numbers are never signed (and do not include 0). It will be seen that this change of viewpoint is of no noticeable importance (cf. footnote 9).

of us were first introduced to the topic, and it does seem a particularly easy one.

Now if we consider a real number written as an unending decimal there is an obvious way in which this number is determined by a sequence[2] of rational numbers, namely the sequence of ending decimals which are the successive initial segments of the unending decimal. For example the positive real number which is a square root of 2 has an unending decimal expansion beginning

$$1 \cdot 4142 \ldots$$

and on this conception we regard it as determined by the unending (and ever-increasing) sequence of ending decimals

$$1, \ 1 \cdot 4, \ 1 \cdot 41, \ 1 \cdot 414, \ 1 \cdot 4142, \ldots$$

Similarly the negative square root of 2 will be determined by the unending (and ever-decreasing) sequence of ending decimals

$$-1, \ -1 \cdot 4, \ -1 \cdot 41, \ -1 \cdot 414, \ -1 \cdot 4142, \ldots .$$

The real numbers themselves can be thought of as the *limits* of these sequences, or alternatively we may think of the first as the *least upper bound* of the increasing sequence and of the second as the *greatest lower bound* of the decreasing sequence. In these examples it evidently makes no difference which way we prefer to look at it, though the idea of a limit to the sequence may lead us to generalize in the direction of Cantor's analysis of the real numbers, while the idea of a least upper bound or greatest lower bound suggests a generalization more in harmony with Dedekind's approach[3]. It is certainly likely to strike one that some generalization would be desirable, since it seems very parochial to concentrate exclusively on the *decimal* notation. There is surely no special magic in the number 10.

One evident generalization of the approach would be to

---

[2] Throughout this section I conform to what seems to be the usual mathematical practice, and use a *sequence* to mean any series which has the general form of a progression (i.e. with a first term, for each term a next, and no others) except that repetitions are permitted.

[3] I take Cantor and Dedekind as the leading exponents of what I am calling the 'classical' approach to real numbers (see Cantor 1883, Dedekind 1872, or almost any modern introduction to Analysis).

suggest that perhaps *any* increasing or decreasing sequence of rational numbers could be used to pick out a real number, but a moment's reflection shows that this would be a mistake. For example it is quite unclear how the sequence of all positive integers (integral rationals) taken in their natural order could be seen as picking out some real number, because that sequence has no limit and its terms have no least upper bound. We must require that our increasing sequences should have limits or least upper bounds and similarly that our decreasing sequences should have limits or greatest lower bounds. However, it may then strike us that what is important is just that we are dealing with a sequence which has a limit, and it does not matter whether it is an increasing or a decreasing sequence, or any other kind of sequence, so long as it has a limit. Or it may strike us that what is important is just that we are dealing with a set of rational numbers that has a least upper bound or greatest lower bound, and it does not matter whether they are thought of as falling into any kind of sequence. So here are two different generalizations of the approach. Of course we can hardly assume without more ado that these different generalizations will coincide with one another, in the sense of yielding the same real numbers with the same properties, or that they will coincide with our first approach in terms of unending decimals.

Let us begin by elaborating the approach in terms of limits of sequences, which is due to Cantor. First, what does it mean to say that a sequence has a limit? Intuitively the basic idea is that as one progresses further and further along the sequence one approaches nearer and nearer to the desired limit, so that the difference between each term and the limit grows progressively smaller and smaller until it becomes less than any difference you may care to specify, however minute. But we can widen this conception slightly. It is not necessary to suppose that at each succeeding term of the sequence the difference from the limit grows less; aberrations from the smooth approach to the limit may be tolerated so long as the aberrations themselves grow less and less. This leads us to the idea behind Cauchy's definition: a sequence converges to a limit N provided that for any difference ε that may be specified (however small) after some point in the series (however late) *all* succeeding terms are within ε of N. The differences ε may obviously be

Infinitesimals ?

taken to be positive rational numbers (since no difference can be less than *every* positive rational number), and therefore we can give a criterion for a sequence having a limit without mentioning the limit itself and without mentioning real numbers at all. We say that a sequence $\langle r_i \rangle$ of rational numbers *has a limit* (or *is convergent*) if and only if for any positive rational ε there exists a term of the sequence r_n such that the difference between r_n and any later term is less than ε, i.e. $(\forall \varepsilon)(\exists n)(\forall m)(|r_n - r_{n+m}| < \varepsilon)$[4]. We can also give a criterion for when two convergent sequences $\langle r_i \rangle$ and $\langle s_i \rangle$ should be said to *have the same limit* again without mentioning what that limit is and without mentioning real numbers at all, namely that when the two sequences are compared term by term then for any positive rational number ε (however small) there should be some point in the sequence (however late) after which the difference between corresponding terms is always less than ε, i.e. $(\forall \varepsilon)(\exists n)(\forall m)(|r_{n+m} - s_{n+m}| < \varepsilon)$. For convenience we may write $\langle r_i \rangle \equiv \langle s_i \rangle$ to mean that the convergent sequences $\langle r_i \rangle$ and $\langle s_i \rangle$ have the same limit, and $[\langle r_i \rangle]$ for the equivalence class of all convergent sequences that have the same limit as the sequence $\langle r_i \rangle$. Clearly on this conception every real number will correspond to a class $[\langle r_i \rangle]$, and conversely, so we may either *identify* the real number with this class (which was Cantor's way) or we may *postulate* that there are real numbers to correspond to each of these classes.

The usual relations and operations on real numbers are then introduced by defining relations and operations on convergent sequences and thereby on their equivalence classes. For instance, if $\langle r_i \rangle$ and $\langle s_i \rangle$ are any convergent sequences of rationals we introduce $\langle r_i \rangle + \langle s_i \rangle$ as the sequence $\langle r_i + s_i \rangle$ whose n^{th} term, for each n, is $r_n + s_n$, and this is easily shown to be again a convergent sequence. Also we may show that if $\langle r_i \rangle \equiv \langle t_i \rangle$ then $\langle r_i \rangle + \langle s_i \rangle \equiv \langle t_i \rangle + \langle s_i \rangle$, so the sum of two real numbers will be given by the sum $[\langle r_i \rangle] + [\langle s_i \rangle]$ of the two corresponding equivalence classes, which in turn is simply $[\langle r_i \rangle + \langle s_i \rangle]$. Multiplication may be defined analogously. To

---

[4] This is Cauchy's criterion. Cantor himself used a slightly different (but equivalent) criterion: a sequence converges if and only if for any positive rational ε all the terms except a finite number differ from one another by less than ε.

define inequalities we first say that a convergent sequence $\langle t_i \rangle$ *has a positive limit* if and only if there is some positive rational number ε such that after some point in the sequence all terms are greater than ε, i.e. $(\exists \varepsilon)(\exists n)(\forall m)(t_{n+m} > \varepsilon)$, and then we define $[\langle r_i \rangle] > [\langle s_i \rangle]$ to mean that there is some convergent sequence $\langle t_i \rangle$ with a positive limit such that $[\langle r_i \rangle] \equiv [\langle s_i \rangle] + [\langle t_i \rangle]$. It is then quite a straightforward matter to develop the arithmetic of real numbers from this basis.

Now let us see how this Cantorian approach to real numbers relates to our other approaches. First, it is clear that every real number that can be written as unending decimal will appear as one of Cantor's real numbers, since as we saw the decimal notation can itself be viewed as a notation for convergent sequences. Next we show that every one of Cantor's real numbers can be viewed as the least upper bound or greatest lower bound of a set of rational numbers, and for this we must establish a lemma. Given any convergent sequence, let us say that the *low* terms of the sequence are those terms (if any) such that no later terms are less than they, and the *high* terms of the sequence are those terms (if any) such that no later terms are greater than they. Then the lemma is that every convergent sequence either contains an infinity of low terms or contains an infinity of high terms (and perhaps it contains both). For suppose a sequence $\langle r_i \rangle$ contains only finitely many low terms and only finitely many high terms. Then there must be some term of the sequence, say r_m, which has all the low terms and all the high terms occurring before it. Since r_m is neither a low nor a high term there occurs after it some term, say r_p, which is greater than it and some term, say r_q, which is less. Further, since r_p is not a high term, there must be a term after it which is greater than it, and since *that* term is not a high term there must be a term after *it* which is greater than it, and so on indefinitely. Therefore terms greater than r_p will never all be left behind, and similarly, since neither r_q nor any of its successors are low terms, terms less than r_q will never all be left behind. But now it is easy to see that the sequence cannot converge, for we have shown that for every term r_n in the sequence there will be a later term greater than r_p and a later term less than r_q, and therefore the difference between r_n and all later terms cannot be less than $(r_p - r_q)/2$, however late in the sequence r_n is taken.

This evidently contradicts the definition of convergence. Thus we establish our lemma: every sequence which does converge contains either an infinity of low terms or an infinity of high terms.

The next point is that if we take any converging sequence $\langle r_i \rangle$ and select from it an infinite sub-sequence $\langle s_i \rangle$, then the sub-sequence $\langle s_i \rangle$ must converge to the same limit as does $\langle r_i \rangle$. For consider any positive rational ε. Now we know that since $\langle r_i \rangle$ converges there is some term r_n such that the difference between it and all later terms is less than $\varepsilon/2$. Hence between any terms of $\langle r_i \rangle$ occurring after r_n the difference must be less than ε, i.e. $(\forall pq)(|r_{n+p} - r_{n+q}| < \varepsilon)$. However, since $\langle s_i \rangle$ is a sub-sequence of $\langle r_i \rangle$, any term s_{n+p} must also be a term r_{n+q} of the sequence $\langle r_i \rangle$ occurring after the nth term of $\langle r_i \rangle$, i.e. $(\forall p)(\exists q)(s_{n+p} = r_{n+q})$. It follows at once that $(\forall p)(|r_{n+p} - s_{n+p}| < \varepsilon)$, which is to say that the two sequences converge to the same limit.

Applying this argument to the preceding lemma, we see that from any converging sequence we can select either a sub-sequence consisting wholly of low terms or a sub-sequence consisting wholly of high terms, and these sub-sequences must converge to the same limit as the original. Therefore without any loss of generality we may in fact confine our attention to sequences composed wholly of low terms or composed wholly of high terms, i.e. to sequences which never decrease or never increase. This gives us our connection with the approach in terms of least upper bounds and greatest lower bounds, for it is intuitively obvious that the limit of a non-decreasing sequence must be the same as its least upper bound and the limit of a non-increasing sequence must be the same as its greatest lower bound. Of course we cannot strictly *prove* this without assuming either that there do exist real numbers to be the required limits or that there do exist real numbers to be the required least and greatest bounds (and we should have to assume or prove that these real numbers have the appropriate properties[5]). What we

[5] Supposing that the definition of $N = \lim\langle r_i \rangle$ is rephrased so as not to presuppose that a real and a rational number can be subtracted from one another, for example by defining it as

$$(\forall \varepsilon_{>0})(\exists n)(\forall m)(r_{n+m} - \varepsilon < N < r_{n+m} + \varepsilon)$$

the proofs would require only a knowledge of rational arithmetic and postulates (1), (5), and (6) below (pp. 205–6).

could do is to show without any further assumption that at least the identity holds when the limits or bounds in question happen to be rational, but I pass over the proofs since, as I say, the propositions to be proved do seem very obvious to intuition. Let us take it, then, that this is sufficient to show that every one of Cantor's real numbers, i.e. every limit to a converging sequence of rational numbers, can be represented equally as a least upper or greatest lower bound of a set of rational numbers.

Further we may without loss confine our attention to least upper bounds in preference to greatest lower bounds (or *vice versa*), since every real number which can be represented as the greatest lower bound of a set α of rationals can also be represented as the least upper bound of the set β of all rationals which are less than every rational in α. Again we cannot strictly *prove* this without assuming something about the real numbers which are supposed to be our least upper and greatest lower bounds. In this case the relevant assumption is that any two distinct real numbers are separated by a rational number, and that any rational number is (or at least corresponds to) a real number[6], and with this assumption the argument might be put thus. Assume that α and β are sets of rationals as above, and suppose first that α contains a least number r. Then r is the greatest lower bound of α, and is the least rational not in β. However, if α contains a least member than β will not contain a greatest (owing to the density of the rationals), so r is the least rational greater than or equal to every member of β. Hence it follows that r is (or corresponds to) the real number which is the least upper bound of β. For if r were greater than some real number N which in turn was greater than all the rationals in β, then by our assumption r and N would be separated by a further rational, and this further rational would therefore be less than r but greater than all the rationals in β, contradicting our conclusion that r was the least rational greater than all the rationals in β. We deduce that in this case r, which is the greatest lower bound of α, is also the least upper bound of β. The next case is that in which β contains a greatest member, and in this case we show by a precisely similar argument that the

[6] Postulates (1) and (6c) below (pp. 205–6).

greatest member of β is the least upper bound of β and also the greatest lower bound of α. In the remaining case α contains no least member and β contains no greatest member. It is then obvious that the greatest lower bound of α must be greater than all members of β and so cannot be less than the least upper bound of β. But also it cannot be greater, for if it were then by our assumption there would be a rational separating them and this rational would be less than every member of α but also greater than every member of β, contradicting the hypothesis that β contains all rationals less than every member of α. Therefore in this case too the greatest lower bound of α must also be the least upper bound of β, and this completes the argument.

Of course it may be asked: Why *should* we make the assumption that any two distinct real numbers are separated by a rational? (Also, why should we make the assumptions needed to show formally that the limit of a non-decreasing sequence is also its least upper bound?) At this stage the only answer we can give is that the required assumptions will be a consequence of our adopting any one of the approaches to real numbers so far canvassed, i.e. the approach in terms of unending decimals, of limits to sequences, of greatest lower bounds, or of least upper bounds. Whichever of these approaches is adopted, the required assumptions will be deducible from the axioms or definitions that we set up in pursuing that approach. At a later stage in the enquiry (§3), I think we shall be able to give a better answer. For the moment, however, let us continue with the approach in terms of greatest lower or least upper bounds, which as we have just seen can be confined without loss to the consideration of least upper bounds.

Parallelling the Cantorian conception, we could think of the real numbers as answering to equivalence classes of non-empty bounded sets of rational numbers, the equivalence relation between any two such sets being that they have the same least upper bound. Further we could explain what it means to have the same least upper bound, without yet supposing that there are any real numbers to be these bounds, by stipulating that two sets of rational numbers are to be said to have the same least upper bound if and only if they have all their rational upper bounds in common. (That is, if α and β be two sets of rationals, we write α ≡ β whenever every rational which is greater than or equal to all members of α is also greater than or equal to all members of β, and conversely.) Dedekind himself did not choose this course, but singled out some preferred sets of rational numbers called *cuts*. A cut is in effect a set which contains *all* the rational numbers less than some real number which is their least upper bound, and it may be defined (without mentioning real numbers) as a non-empty set of rationals such that (i) there are some rationals not in the set, (ii) if any rational is in the set then all rationals less than it are in the set, and (iii) if any rational is in the set then some rational greater than it is in the set[7]. It is easy to show from the suggested criterion for having the same least upper bound that to every non-empty bounded set of rationals there must correspond a cut which has the same least upper bound, *viz.* the set of all rationals less than some member of the original set, so no generality is being lost if we once more restrict our attention to sets of rationals which are cuts. The restriction simplifies matters because it is obvious that distinct cuts cannot have the same least upper bounds, so there is just one cut to each real number. Hence we may say (as Dedekind himself said) that to each distinct cut there somehow *corresponds* a distinct real number which is determined by it, or we may follow Cantor's procedure and *identify* the real number with the cut that determines it. In either case we shall introduce relations and operations on real numbers via relations and operations on their corresponding cuts. For example the sum of two real numbers will be given by the sum $\alpha + \beta$ of their corresponding cuts, which in turn is the cut consisting of all sums $r + s$ of rationals r and s such that r is in α and s is in β, and similarly in other cases.

Now we may return to our task of showing that these various conceptions of real numbers coincide. So far we have shown that every real number conceived as an unending decimal may equally be conceived as the limit of a convergent sequence of rational numbers, that every real number conceived as the limit

---

[7] In Dedekind's own formulation a cut is rather a *pair* of sets (α, β) such that *either* α is a cut in our sense and β is the set of all rationals not in α, *or* β is the dual to a cut in our sense (with 'less than' and 'greater than' interchanged in the definition) and α is the set of all rationals not in β. Equivalently, if we drop clause (iii) from the definition of a cut, then α is to be a cut in this looser sense and β is to be its complement. (For the genesis of Dedekind's idea see p. 218.)

of a convergent sequence may equally be conceived as a greatest
lower bound or least upper bound of a set of rational numbers,
and finally that every real number conceived in this latter way
may in particular be conceived as the least upper bound of a cut.
To complete the circle we may now show that every real number
conceived as the least upper bound of a cut may also be conceived as an unending decimal, i.e. that to every cut there
corresponds an appropriate sequence of ending decimals with
the same least upper bound. Once more we must distinguish
cases. Suppose first that the least upper bound of the cut is
positive, i.e. that the cut contains some positive rational. Now
we establish a lemma: every positive member of the cut is less
than some rational which can be expressed as an ending decimal and which is itself a member of the cut. To see this, we
have only to consider the decimal expansion of rational numbers. Let r be any positive member of the cut. Then (by the
definition of a cut) there is another member s of the cut which is
greater. When r and s are expressed in decimal notation there
must be a digit in which they differ, and the shortest initial
segment of the expansion of s which differs from the corresponding initial segment of the expansion of r will be an ending
decimal greater than r and less than or equal to s. Since it is
less than or equal to s, and s is a member of the cut, then (by
the definition of a cut) it must itself be a member of the cut. So
this proves the lemma, and from it the desired conclusion follows
without difficulty. For suppose we select from our cut first the
greatest integer in the cut (which may be 0 but by hypothesis
cannot be less), then the greatest ending decimal with one digit
after the decimal point, then the greatest ending decimal with
two digits after the decimal point, and so on. Clearly this must
be an unending sequence of ending decimals so arranged that
it allows us to construct an unending decimal from it. Also,
since it is selected from the cut, its least upper bound cannot be greater than that of the cut. Equally, its least upper
bound cannot be less than that of the cut, since every member
of the cut is of course less than some positive member, and
every positive member is (by the lemma) less than some member of the sequence. So this establishes the desired conclusion for
the case when the least upper bound of the cut is positive. In
the case when the least upper bound of the cut is negative, which

is to say that there is some negative rational *not* in the cut, we proceed by parallel considerations to select a decreasing (or rather, non-increasing) sequence of ending decimals from the negative rationals that are not in the cut. The unending decimal constructed from it is clearly the greatest lower bound of the rationals that are not in the cut, and therefore the least upper bound of the rationals that are in the cut. The only remaining case is that in which the least upper bound of the cut is zero, and this case is trivial.

This completes the argument to show that all the suggested approaches to real numbers are equivalent, in the sense that if we think of a real number as determined by, say, a convergent sequence of rational numbers as its limit then that real number is equally determined by, say, a non-empty bounded set of rational numbers as its least upper bound, and conversely. So from the point of view of *postulating* real numbers, it will make no difference whether we postulate that limits always exist or that least bounds always exist, for either postulate implies the other. If we wish rather to adopt a 'logical construction' approach then of course the various constructions are equivalent only in a looser sense, for one cannot consistently *identify* one and the same real number *both* with an equivalence class of convergent sequences *and* with an equivalence class of non-empty bounded sets, since these are different classes. However, by an argument which is now very familiar, this point simply destroys the logical construction approach: since both identifications are *equally* justifiable, neither can be *adequately* justified, so both should be rejected. Nor does it help to bypass the identification (as suggested for rational numbers on p. 79) and rewrite whole statements concerning real numbers as statements directly concerning convergent sequences of rationals, or non-empty bounded sets of rationals, or cuts, or whatever. For again the problem arises as to *which* of the various available reductions is *correct*, i.e. which gives the right account of the meaning of statements apparently concerning real numbers. For example on Dedekind's lines the statement that for any real numbers N and M either $N \leqslant M$ or $M \leqslant N$ would be identified with the statement that for all cuts α and β of rational numbers either $\alpha \subseteq \beta$ or $\beta \subseteq \alpha$. On Cantor's approach its meaning would be much more complicated: roughly, for all convergent sequence $\langle r_i \rangle$ and $\langle s_i \rangle$, either $\langle r_i \rangle \equiv \langle s_i \rangle$ or there is a convergent sequence $\langle t_i \rangle$ and a positive rational ε such that $(\exists n)(\forall m)(t_{n+m} > \varepsilon)$ and either $\langle r_i \rangle \equiv \langle s_i \rangle + \langle t_i \rangle$ or $\langle r_i \rangle + \langle t_i \rangle \equiv \langle s_i \rangle$. It is surely grossly implausible to maintain that these two 'analyses' mean the same as one another, so at least one must be incorrect. Therefore both are incorrect unless some further considerations can be adduced for preferring one to the other. However, without going beyond the classical 'arithmetizing' approach, whereby real numbers must some-how be explained in terms of other numbers already available, it would not seem that there are any relevant further considera-tions.

Perhaps someone might invoke *simplicity*, for it does seem to be the case that in practice an analysis on Dedekind's lines is seldom more complicated than one on Cantor's, and often, as in the above example, simpler. However, it is not clear to me. why simplicity should be a relevant consideration, and besides our ordinary concept of simplicity is probably too vague to allow us to pick just one of these patterns of 'analysis' as the simplest. But anyway, suppose we grant for the sake of argument that Dedekind cuts yield the simplest 'analysis' so far con-sidered. Still it does have one unnecessary complexity which every such logicist account is bound to have, and that is that on these accounts real numbers must be a quite different type of thing from rational numbers. (Since it cannot make sense to identify a rational number with the set of all rationals less than it—for how in *that* case could one explain the underlying rational arithmetic?—no rational number can actually be a cut, but every real number is a cut.) But this means that the operations of, for example, adding or multiplying a real num-ber and a rational number must strictly speaking be operations on entities of different types, and similarly for the relations (such as being less than) which we ordinarily think of as hold-ing between real and rational numbers. From the point of view of our ordinary conception this seems a quite gratuitous com-plication, and so I think that even if simplicity did rule in favour of one logicist account rather than any other, still every logicist account will seem less simple than the postulational approach which leaves real numbers and rational numbers as entities of the same type. Let us turn then to consider in more

detail how real numbers may be not constructed but postulated[8].

When considering the rational numbers we began by pointing
out that although any two natural numbers could be multiplied,
the inverse operation of division was not always performable,
and the rational numbers were then seen as *added* to the natural
numbers in order to ensure that this operation could always be
carried out. An elegant feature of the resulting system was that
although we began by requiring only that division could always
be applied to natural numbers, and this then led us to introduce
our new rational numbers, it turned out (p. 80) that in the
new system division could *also* be applied to any two rational
numbers. So we were able to close this gap (as it were) in the
natural numbers without at the same time having that same
gap reappear in the new rational numbers. But we did open up
a different gap. It is an important feature of the system of
natural numbers that every non-empty set of natural numbers
which has an upper bound has a least upper bound. This
feature is lost when we extend the natural numbers to the
rational numbers, and the attempt to regain it will lead us
on, as we have seen, from the rational numbers to the real
numbers. So I choose my postulates to conform to this approach[9].

Let us suppose that r, s, t, . . . are variables already in
use ranging over rational numbers, and that Φ, Ψ, . . . are
variables for predicates of rational numbers. We shall use N,
M, . . . as variables for real numbers, and our postulates will
fix the range of these variables. Now the idea is that the real
numbers shall be precisely the least upper bounds of non-
empty and bounded sets of rational numbers. It follows at once

---

[8] I have been identifying 'logical construction' accounts with logicist
accounts, but I hesitate to identify the contrasting postulations with formalist
accounts. It is characteristic of the postulates that I am presently considering
that they *relate* real numbers to rational numbers, as earlier rational numbers
were related to natural numbers. Many formalists would prefer to start afresh
with an entirely new set of axioms for real numbers, but that approach seems
to me of no interest.

[9] The approach is also well suited to the present book, for henceforth it will
make no difference to the discussion whether we take our rational and real
numbers to be signed (and therefore to include 0), or whether we restrict
attention to positive numbers only.

that every rational number is a real number, and so we have our first postulate:

(1) $(\forall r)(\exists N)(r = N)$.

Further, since every real number is a least upper bound of the required sort we have our second postulate:

(2) $(\forall N)(\exists \Phi)((\exists r)(\Phi r) \ \& \ (\exists s)(\forall r)(\Phi r \to r \leqslant s)$
    $\& \ (\forall s)(N \leqslant s \leftrightarrow (\forall r)(\Phi r \to r \leqslant s)))$.

Conversely, since every such least upper bound is a real number, and we are now postulating that the desired least upper bounds always exist, we have our third postulate affirming existence:

(3) $(\forall \Phi)((\exists r)(\Phi r) \ \& \ (\exists s)(\forall r)(\Phi r \to r \leqslant s)$
    $\to (\exists N)(\forall s)(N \leqslant s \leftrightarrow (\forall r)(\Phi r \to r \leqslant s)))$

To this we must add a fourth affirming uniqueness, i.e. if N and M are least upper bounds of sets of rationals which have all their rational upper bounds in common, then N and M must be the same real number. This is most simply secured by laying down

(4) $(\forall s)(N \leqslant s \leftrightarrow M \leqslant s) \to N = M$.

We can consider (1)–(4) as the fundamental postulates for this approach, or if preferred we can give an explicit definition of the restricted variable N and deduce (1) and (2) as immediate consequences (as on p. 83).

The postulates (1)–(4) do not yet form an adequate set, for notice that these postulates make use of a relation $\leqslant$ which holds not only between rational numbers but also between other real numbers (or at any rate between a real number and a rational number); indeed one cannot formulate the idea that a real number is a least upper bound without employing such a relation. However, the properties of this relation are not wholly determined by (1)–(4). We may say that just as in the case of rational numbers our fundamental postulates presupposed that we knew how to multiply a rational and a natural number together, at least in certain cases, so here the fundamental postulates presuppose that we know how to compare a rational and a real number in magnitude. Again, just as in the previous case we had to go on and postulate that multiplication of arbitrary rational numbers preserved certain properties, so here too we must lay down something further about the relation $\leqslant$. A natural stipulation would be that $\leqslant$ remains a totally ordering relation, which we would secure by laying down[10]

(5a) $N \leqslant M \lor N \geqslant M$

(5b) $N \leqslant M \,\&\, N \geqslant M \to N = M$

(5c) $N \leqslant M \,\&\, M \leqslant P \to N \leqslant P.$

(This covers 'mixed' cases in view of (1)). A set of postulates equivalent to (1)–(5) and slightly neater is obtained by strengthening (4) and weakening (5c) to obtain the equivalence[11]

(4*) $N \leqslant M \leftrightarrow (\forall s)(M \leqslant s \to N \leqslant s).$

The postulates (1)–(4*) suffice for the deduction of all theses in (5) and for such further properties of $\leqslant$ as[12]

(6a) $(\exists r)(r < N)$

(6b) $(\exists r)(N < r)$

(6c) $N < M \to (\exists r)(N < r \,\&\, r < M).$

They also suffice to prove the theorem which shows that our attempt to 'complete' the system of rational arithmetic by providing least upper bounds for all appropriate sets of rational numbers has *also* succeeded in providing least upper bounds for all appropriate sets of real numbers, i.e.

(7) $(\exists N)(\Phi N) \,\&\, (\exists M)(\forall N)(\Phi N \to N \leqslant M)$

$\quad \to (\exists P)(\forall M)(P \leqslant M \leftrightarrow (\forall N)(\Phi N \to N \leqslant M)).$

The proof of this theorem proceeds on entirely orthodox lines, but since the theorem is of some importance it may be well to give a quick sketch of it.

---

[10] Postulate (5b) is superfluous in view of (5c), (1), and (4).

[11] Note that this equivalence cannot be treated as a definition, since $N \leqslant s$ is a special case of $N \leqslant M$.

[12] I assume that $<$ is defined in terms of $\leqslant$ in the standard way.

We begin, then, by assuming the antecedent

(i) $(\exists N)(\Phi N)$ & $(\exists M)(\forall N)(\Phi N \to N \leqslant M)$.

Applying (6a) to the first conjunct here and (6b) to the second (and using (5c)) we easily deduce

(ii) $(\exists r)(\exists N)(r < N$ & $\Phi N)$

    & $(\exists s)(\forall r)((\exists N)(r < N$ & $\Phi N) \to r \leqslant s)$.

Now (ii) constitutes the antecedent for an application of postulate (3), with $(\exists N)(r < N$ & $\Phi N)$ substituted for Φr. Therefore we apply this postulate and deduce that for some real number P

(iii) $(\forall s)(P \leqslant s \leftrightarrow (\forall r)((\exists N)(r < N$ & $\Phi N) \to r \leqslant s))$.

Our task now is to show that P is the least upper bound of the real numbers N such that ΦN. First we show that it is at least an upper bound. Assume

(iv) ΦN.

Then, using (iii) from left to right, we have

    $(\forall s)(P \leqslant s \to (\forall r)(r < N \to r \leqslant s))$

that is

    $\sim(\exists rs)(P \leqslant s$ & $s < r$ & $r < N)$.

Therefore, using (6c) twice,

(v) $\sim(P < N)$.

Conditionalizing the argument from (iv) to (v) we have

(vi) $(\forall N)(\Phi N \to N \leqslant P)$

which is the desired conclusion. Next, to show that P is the least upper bound, assume

(vii) $P > M$.

Then by (6c) we have

    $(\exists s)(P > s$ & $s > M)$.

Therefore, using (iii) from right to left, we have

    $(\exists s)((\exists r)(\exists N)(r < N$ & ΦN & $r > s)$ & $s > M)$

That is

$$(\exists N)(\exists rs)(\Phi N \ \& \ N > r \ \& \ r > s \ \& \ s > M)$$

whence by (5c)

(viii) $(\exists N)(\Phi N \ \& \ N > M)$.

Conditionalizing the argument from (vii) to (viii) and contraposing, we obtain

(ix) $(\forall N)(\Phi N \rightarrow N \leqslant M) \rightarrow P \leqslant M$

which again is the desired conclusion. This completes the proof of (7).

The properties of the relation $\leqslant$ between real numbers are then completely given by postulates (1)–(4*), and no further postulates are required[13]. Further, there is no difficulty in filling out this approach by providing suitable definitions of addition, multiplication, and so on. For example the sum $N + M$ may be defined as the least upper bound of all rationals $r + s$ such that $r \leqslant N$ and $s \leqslant M$, and it can be shown that *only* this definition (or an equivalent) will preserve for real numbers the formal properties of addition which hold for rational and for natural numbers. Thus suppose we stipulate that addition of real numbers must be commutative and must have the property that

(i) $M \leqslant N \leftrightarrow M + P \leqslant N + P$.

Then one special case covered by this requirement is

(ii) $r \leqslant N \leftrightarrow r + p \leqslant N + p$

and from this we may argue that the sum of a real number and a rational number must be as described. For (a) we know from rational arithmetic that

$$s \leqslant p \rightarrow r + s \leqslant r + p$$

and putting this together with (ii) we have

$$r \leqslant N \ \& \ s \leqslant p \rightarrow r + s \leqslant N + p$$

which is to say that $N + p$ is *an* upper bound of the rationals

[13] It is well known that (1), (5), (6), and (7) are categorical. The proof was first given by Cantor in 1895.

$r + s$ such that $r \leqslant N$ and $s \leqslant p$. To show (b) that it is the least upper bound, we note that if

$$(\forall rs)(r \leqslant N \ \& \ s \leqslant p \rightarrow r + s \leqslant M)$$

then clearly

$$(\forall r)(r \leqslant N \rightarrow r + p \leqslant M)$$

and so, using (ii) from right to left,

$$(\forall r)(r + p \leqslant N + p \rightarrow r + p \leqslant M)$$

From this it is easy to see, by a simple extension of (4*), that

$$N + p \leqslant M$$

which shows that $N + p$ is the least upper bound of these rationals, and so establishes our conclusion at least for sums of a real and a rational number. Using this result, we then extend our argument to cover all sums of real numbers by arguing in the same way from the premise

(iii) $p \leqslant M \leftrightarrow N + p \leqslant N + M$

which is another special case of the leading principle (i). Multiplication may obviously be treated similarly, and so may exponentiation. So no further difficulties are encountered in developing the arithmetic of real numbers from this basis.

This way of approaching the theory of real numbers by postulating the existence of least upper bounds has a certain naturalness, as we have seen. But of course the question arises: Is there any reason to suppose that the postulates are true? By making use of the corresponding logical construction approach we may certainly show that the postulates are consistent provided that the logic used in the logical construction is itself consistent, though in this case the proviso is not altogether negligible. (If the construction is carried out as a part of set theory, with real numbers taken to *be* sets of rational numbers, it is easy to complete the proofs of (1)–(7) without going beyond, say, Zermelo's set theory, but of course the consistency of that theory is not a trivial question. Alternatively if we translate our set theory into higher-order predicate calculus, in accordance with the methods of Russell's theory of types, then we see that the logic must be taken to be impredicative for the proof of (7), and that again makes the question of consistency non-trivial.)

But anyway it does not seem to me that a mere consistency proof would settle our doubts about the approach via postulates, for our original question was whether there was any reason to suppose the postulates *true*, and that question is surely not answered by a proof that they are at least consistent.

But perhaps the question should itself be rejected, for what does the demand for truth come to in this case? It must evidently presuppose that we have some independent understanding of the real numbers, which could provide a check on the postulates, for otherwise there is no possible way of attacking the question. Now it may be said that in this section we have already suggested some independent ways of understanding real numbers, for example as the numbers which can be expressed as unending decimals, and we have already sketched an argument to show that the postulates are justified on the basis of this conception. But again, is there anything which makes us think that an unending decimal does always represent a number? Is there any *need* for such numbers, or should we conclude that they are (roughly speaking) just an elegant creation of the mathematician's invention? When the question is put in this way it obviously invites us to consider what we now class as the *applications* of the theory of real numbers, and there is no doubt that in the historical development of this theory one area of 'application' was of fundamental importance, namely geometry.

Consider, for example, the classic case of the square root of 2. There is no rational number which is a square root of 2, as the Greeks very soon discovered, so why do we not say that there simply is *no* number which is a square root of 2, just as we do generally say that there is no number which is the result of dividing 2 by 0? Now if we turn to elementary geometry an answer seems to be provided, for geometry appears to demand such a number. For example we may easily show (say by considering the diagram on the next page) that if a larger square ABCD has as its side the diagonal of a smaller square AEDF, then the larger square has an area which is twice that of the smaller. So if the smaller square has a side of 1 inch long, and therefore an area of 1 square inch, the larger square will have an area of 2 square inches. From this it appears to *follow* that the side of the larger square must be $\sqrt{2}$ inches long, for generally a square with its

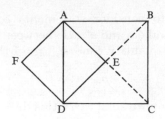

sides n inches long must have an area of n^2 square inches. This is, however, moving a little too fast, for more strictly the conclusion does not follow without further assumption. What does follow is that the diagonal AD is more than r inches long for any rational number r such that $r^2 < 2$, and less than r inches long for any rational number r such that $r^2 > 2$. We only obtain our conclusion by adding the further requirement that there must be some number N such that the diagonal is exactly N inches long (whence $N^2 = 2$, so $N = \sqrt{2}$). This further assumption, that every line must have a definite length in inches, could perfectly well be denied—classical Greek mathematics did in fact deny it—but it is surely this sort of assumption that lies behind our conviction that there is such a number as $\sqrt{2}$.

To put the suggestion more generally, it is that irrational numbers are needed to provide numerical measures for the various degrees of continuous quantities (such as length, area, mass, time, and many others), and it seems clear that as a matter of history this need became felt long before any need for, say, limits of convergent sequences of rational numbers or least upper bounds of bounded sets of rational numbers. To provide a better understanding of the character of this need I propose now to consider how the study of continuous quantities could proceed without assuming the existence of irrational numbers, for classical Greek geometry was of course pursued without this assumption.

2. Geometry and the Theory of Proportion

In the last section of Chapter 2 we considered statements comparing two ratios between natural numbers, for example of the form

$$n:m \geqslant p:q$$

and we went on to consider statements comparing a ratio between natural numbers and a ratio between two other objects (in respect of some quantity φ-ness), e.g. of the form

$$a:b \geqslant n:m.$$

In the present section we shall carry this line of thought further and introduce the general theory of proportion which permits us to compare two ratios between things of any sort and so contains statements of the form

$$a:b \geqslant c:d.$$

The first problem is evidently to determine when we should say that the ratio of a to b and the ratio of c to d (in respect of some quantity) are the *same* ratio.

Now when we were considering ratios just between natural numbers we said that the fundamental idea was that the ratio of n to m should be the same as that of p to q when *all* relations of numerical comparison which held between n and m also held between p and q (and conversely), but we could not have used this criterion as a definition because at the time the relations of numerical comparison available to us were limited to those employing natural numbers only (i.e. of the form $\geqslant_n$, $\leqslant_n$, and $\approx_n$). Consequently we relied instead on a special feature of the natural numbers, that they are all commensurable—i.e. each is a definite fraction of every other—because it is obvious that if n is the same fraction of m as p is of q then they must also share all relations of numerical comparison. We now wish to give a more general criterion for sameness of ratio and can no longer rely on commensurability, but to compensate for this we do now have a greater stock of relations of numerical comparison available, *viz.* all those employing rational numbers. These are not indeed the only relations of numerical comparison, for we have yet to introduce real numbers in this context, but nevertheless it turns out (as I shall argue in a moment) that if a and b share all *these* relations of numerical comparison with c and d then they must also share all other relations of numerical comparison. Further, since all relations of numerical comparison employing rational numbers may be defined in terms of $\geqslant_{(n/m)}$, the suggested criterion could be put as the claim that the ratios of a to b and of c to d are the same ratio if and only if

$$(\forall nm)(a \succcurlyeq_{(n/m)} b \leftrightarrow c \succcurlyeq_{(n/m)} d).$$

But it needs to be shown that this criterion is adequate. It is of course obvious that the suggested criterion is a necessary condition of sameness of ratio, and the problem—to which I now turn—is to show that it is also a sufficient condition.

We may observe first that no difficulty arises if a and b (or c and d) are commensurable, for since the suggested criterion entails

$$(\forall nm)(a \approx_{(n/m)} b \leftrightarrow c \approx_{(n/m)} d)$$

it is easily seen that if the ratios of a to b and of c to d are equal according to this criterion then, if a is some fraction of b, c must also be the same fraction of d, and conversely. So if there is any difficulty it must arise in the incommensurable case, and therefore where the quantity concerned is continuous. Now the crucial fact is that we can prove that if the ratios of a to b and of c to b are equal according to our criterion, then a and c must themselves be equal, i.e. we can prove that

$$(\forall nm)(a \succcurlyeq_{(n/m)} b \leftrightarrow c \succcurlyeq_{(n/m)} b) \rightarrow a \approx c$$

and this holds whether or not a and b (or c and b) are commensurable. However, it is clear that if a and c are equal then they must share *all* relations of numerical comparison to any third thing b, and therefore that the suggested criterion must be sufficient in this special case. Hence we can argue to its more general sufficiency. Suppose that we have two ratios, that of a to b and that of c to d, which are equal according to our criterion, and suppose for example $b \succcurlyeq d$. (The case where $d \succcurlyeq b$ may be treated symmetrically). Now if $b \succcurlyeq d$, and the quantity concerned is continuous, it is, I take it, intuitively evident that when we mark off a part y of b equal to d we can find a proportionate part x of a such that the ratio of a to b is the same as that of x to y, and therefore (since $y \approx d$) the same as that of x to d. Of course we can give a formal proof of this claim once the criterion has been adopted as a definition, i.e. we can prove from the definition that for continuous quantities

$$b \succcurlyeq d \rightarrow (\exists x)(a : b = x : d).$$

But since we are now considering whether that criterion is adequate as a definition, it would be circular to rely here upon a proof which itself relies upon the definition. So my present claim is that this thesis holds for sameness of ratio anyway, whether or not the suggested criterion is adequate[15]. But now, if the ratio of a to b is in fact the same as that of x to d, and is according to our criterion equal to that of c to d, then it follows that the ratios of x to d and of c to d are at least equal according to our criterion. But we began by arguing that in that case the ratio of x to d and of c to d must in fact be the same ratio, since this is the special case first treated of. So we conclude that if the ratios of a to b and of c to d are equal according to our criterion then they must in fact be the same ratio, because (for some x) they are each the same ratio as that of x to d. Hence the suggested criterion is also adequate in this more general case.

In this argument I have been considering only the case in which the statement $a:b=c:d$ concerns the ratio of a to b in respect of some quantity ϕ-ness and the ratio of c to d in respect of *that same quantity*, since I do not think that our intuitions about sameness of ratio give us any firm guidance in the still more general case when the quantities concerned may be different. However, it is now time to observe that this still more general case is perfectly well catered for by our criterion, since there is clearly no need to suppose that in the condition

$$(\forall nm)(a \succcurlyeq_{(n/m)} b \leftrightarrow c \succcurlyeq_{(n/m)} d)$$

the subscripts omitted with each occurrence of $\succcurlyeq_{(n/m)}$ are the same. So we must now restore our subscripts explicitly, since we need no longer assume that the subscripts will be the same

---

[15] It may be objected that this claim relies upon our intuition of continuity, so although the argument may perhaps be admitted for genuinely continuous quantities it is not yet clear that it holds for quantities that are continuous only in the sense that I have so far stipulated (p. 173). Later I shall argue that quantities continuous in the sense stipulated are also continuous in an intuitive sense (pp. 237–9), but *that* argument will presuppose that the present criterion for sameness of ratio is adequate. Therefore it seems that it may yet be a possibility that the present criterion should fail to match our intuitive understanding in the case of quantities which are neither discrete nor *genuinely* continuous. However, there are no such quantities which are relevant, i.e. none which satisfy the axioms, and this is shown by the categoricity of the axioms (pp. 245–8).

throughout any formula we consider. To avoid too great a clutter of subscripts I propose therefore to replace the clause $a \geqslant_{(n/m)} b$ by its definiens $a : b \geqslant n : m$, where the required subscript ϕ can conveniently be added to the first occurrence of :,[16] and I shall then say that the ratio of a to b in ϕ-ness is the same as that of c to d in ψ-ness if and only if

$$(\forall nm)(a :_\varphi b \geqslant n : m \leftrightarrow c :_\psi d \geqslant n : m).$$

For greater generality, however, I do not adopt quite this definition, for we may conveniently lay down at once a more general definition covering inequalities[17]

$$a :_\varphi b \geqslant c :_\psi d \quad \text{for} \quad (\forall nm)(c :_\psi d \geqslant n : m \rightarrow a :_\varphi b \geqslant n : m).$$

The formulae

$$a :_\varphi b = c :_\psi d$$

$$a :_\varphi b > c :_\psi d$$

are then defined in terms of $\geqslant$ in the standard way. The relation $\geqslant$ so introduced is obviously transitive, and we can easily prove that it is also connected, i.e. that

$$a :_\varphi b \geqslant c :_\psi d \lor c :_\psi d \geqslant a :_\varphi b.$$

Hence it follows that our argument to show the adequacy of

---

[16] As observed before (footnote on p. 171) we should strictly speaking subscript not only the comparative relation $\geqslant_\varphi$ but also the associated (quasi-)-operation $\oplus$ of addition.

[17] Since the subscript ψ has as a special case that in which $\geqslant_\psi$ is taken as $\geqslant$ between natural numbers, the formula

$$a :_\varphi b \geqslant c :_\psi d$$

is defined in terms of one of its own special cases

$$a :_\varphi b \geqslant p : q$$

and this in turn was defined in terms of one of its special cases

$$n : m \geqslant p : q$$

So the last of these formulae is now redefined for the third time and the middle one for the second. To avoid this conflict with orthodox requirements on definition we should strictly introduce different signs in place of $\geqslant$ on each occasion and then prove that the relations coincide wherever they are all applicable. However, I continue to ignore such complications (cf. p. 190).

the definition of $=$ carries over without difficulty to the more general definition of $\geqslant$[18].

This completes my introduction of the fundamental definitions of the theory of proportion, but before I go on to develop that theory further I think it will be helpful to see it 'in action'. I therefore propose to illustrate the definitions by considering their application to the historically important case of geometrical quantities. However, as a preliminary it must of course be proved or assumed that geometrical quantities do satisfy our initial axioms, so I now digress to make some remarks on this point.

Early in the development of Euclidean geometry one will certainly introduce the idea of a straight line-segment (or a finite straight line) which will be determined by the two points which are its end-points, and of a congruence-relation defined on these line-segments which will be assumed (or proved) to be an equivalence relation. In the usual way I shall use capital letters A, B, ... to represent points, the concatenation AB for the straight line-segment with A and B as its end-points, and AB$\simeq$CD to signify that AB and CD are congruent. I shall also make use of the idea of one line-segment AC being divided into two others, AB and BC, in such a way that we can reasonably introduce the quasi-term XY$\oplus$ZW as an indefinite description of any line-segment AC, which can be divided into two parts AB and BC, such that AB$\simeq$XY and BC$\simeq$ZW. (It would be entirely natural to use the notion of a *part* explicitly when introducing this idea, but of course it is not essential. For all that is required is that B fall *between* A and C, and betweenness is an essential notion in geometry anyway). Axioms will no doubt be provided sufficient to ensure that

(1) E!(AB$\oplus$CD)

(2) (AB$\oplus$CD)$\simeq$(CD$\oplus$AB)

---

[18] For example we may appeal to the so-called 'law of closed systems' (see e.g. Tarski 1941, § 50). We have argued that the defined $=$ is a sufficient condition for sameness of ratio, and it is easy to see that the defined $>$ (and hence $<$) is sufficient for greater (lesser) ratios. However, we can prove that the defined $=$, $>$, $<$ are exhaustive, and it is obvious that their intuitive counterparts are exclusive. Hence it follows that all three sufficient conditions are also necessary.

(3) $(AB \oplus (CD \oplus EF)) \simeq ((AB \oplus CD) \oplus EF)$

(4) $AB \simeq CD \rightarrow (AB \oplus EF) \simeq (CD \oplus EF)$

(5a) $AB \nleq (AB \oplus CD)$

(5b) $AB \nleq CD \rightarrow$

$$(\exists XY)(AB \simeq (CD \oplus XY) \lor (AB \oplus XY) \simeq CD)$$

In view of the last two theses we may introduce

$$AB \succ CD \quad \text{for} \quad (\exists XY)(AB \simeq (CD \oplus XY))$$

and $\succ$ between line-segments will be a comparative relation with congruence as its associated equality relation. The quantity in question is of course *length*, so to avoid confusion we may conveniently write these relations as $\succ_{\text{length}}$ and $\approx_{\text{length}}$ where distinction is necessary. As postulate (1) requires, it is usually assumed in Euclidean geometry that length is an unbounded quantity, i.e. there is no upper limit to the lengths which line-segments may have, and it will also be assumed that there is no lower limit either. Sooner or later an axiom will be added to ensure that Archimedes' postulate holds

(6) $(\exists n)(AB \prec n \circ CD)$

and no doubt from very early stages it will have been assumed (or proved) that

(7) $(\exists XY)(AB \succ XY).$

From (6) and (7) it follows, as we have seen (pp. 173–4), that there exist line-segments of less than any specified length.

The assumptions so far stated are of course by no means sufficient for Euclidean geometry. Indeed (1)–(7) would usually be deduced from rather stronger axioms which would allow us to establish principles at least as powerful as ADD and COMP (pp. 103–1), but since that is not strictly necessary for our present purpose we may ignore the fact. Besides this, several axioms that are fundamental to Euclidean geometry are entirely independent of (1)–(7), in particular the *transversal* axiom which provides an essential property of the betweenness

relation[19], a further congruence axiom to provide for the properties of congruent triangles, and of course the famous parallel
axiom. When these are added we shall have a basis on which a
great deal of Euclidean geometry can be erected[20], but to complete the basis we shall also have to add a *continuity* axiom
allowing us to deduce that length (of line-segments) is a *continuous quantity*, and therefore one which satisfies the principle
of least upper bounds. The required axiom is usually formulated
in terms of the betweenness relation, roughly as follows:

> If all points of a line-segment are separated into two discrete
> and non-empty sets α and β such that no point in α is between
> any two points in β or *vice versa*, then there is a point X such
> that, if A is any point in α and B any point in β, then either
> X = A or X = B or X is between A and B.

It was by reflecting on this axiom for the line that Dedekind
came to formulate his idea of a *cut*, and of a real number as
determined by a cut in the rational numbers, and it is easy to
see the correspondence between the two ideas[21]. It is also easy
to see that this axiom for the line does indeed imply the principle of least upper bounds for length of line-segments. For
suppose we are given any non-empty set of line-segments and a
line-segment AB which is (in length) an upper bound of the set
but not a least upper bound. Then we can separate the points
of AB into two sets α and β by putting a point C of AB in α
when AC is less than or equal to some line-segment in the
original set, and in β otherwise. This is easily seen to satisfy the
condition on α and β in the axiom, and so we deduce the existence of the 'separating' point X and can show without difficulty that the line AX is a least upper bound of the original

[19] Roughly, if A, X, B and X, O, C occur in that order, then there is a
point Y such that A, Y, C and Y, O, B occur in that order

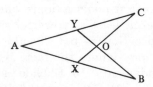

[20] See for example Forder 1927.
[21] See footnote 7, p. 200.

set. (Incidentally the converse deduction is equally straight-forward.) Anyway, as I say, the axioms for Euclidean geometry will almost certainly provide fairly directly for the theses needed to show that *length* (of straight line-segments) is a continuous quantity.

We shall then be able to *prove* that various other quantities occurring in Euclidean geometry are also continuous quanti-ties, for example area and volume. I shall illustrate this with the case of *area* (of a plane figure). The general technique is to find a way of correlating plane figures and line-segments in such a way that every plane figure is correlated with some line-segment and every line-segment is correlated with some plane figure. The correlation need not be one–one (or even one–many or many–one) but must be such that plane figures which are (intuitively) greater, less, or equal in area are correlated with line-segments which are correspondingly greater, less, or equal in length. We may, however, use the correlation itself to pro-vide formal *definitions* of 'greater in area', 'less in area', and 'equal in area', so all that will be formally required is that the *same* plane figure is correlated only with *equal* line-segments. Then where x and y are any plane figures we may define $x \approx y$ to mean that x and y are correlated with line-segments that are equal in length, and we may define $x \oplus y$ as an indefinite des-cription of any plane figure z such that if x is correlated with the line-segment AB and y with CD then z is correlated with AB $\oplus$ CD. The properties of $\approx$ and $\oplus$ between line-segments will then transfer automatically to the new $\approx$ and $\oplus$ between plane figures, and it will follow at once that the quantity of plane figures introduced by this definition is also a continuous quantity. We shall also have to show that the quantity thus introduced is in fact the quantity *area* as we intuitively understand it, and here I shall content myself with showing that the relations $\approx$ and $\prec$ introduced by the definition are equivalent to the relations 'equal in area' and 'less in area' as we understand them.

For this purpose we must agree on some properties of these relations, and besides the obvious facts that both relations are transitive, the first symmetrical and the second asymmetri-cal, I propose to assume here that for any plane figures a, b:

(i) if a is congruent to b, a is equal in area to b;

(ii) if a is a proper part of b, a is less in area than b;

(iii) if a is divided into two plane figures x and y, and if b is divided into two plane figures z and w, then if a is equal in area to b and x is equal in area to z, y is equal in area to w.

We may note incidentally that in Euclidean geometry assumption (iii), that 'equals taken from equals leave equals', implies conversely that 'equals added to equals yield equals'. For, if a and b are any two figures, we may always draw two figures a' and b' such that a' and b' are congruent (and so, by (i), equal in area), and a is a proper part of a' and b a proper part of b'. Hence if a is divided into x and y, and b into z and w, and x and z, y and w are respectively equal in area, then by two 'subtractions' we can deduce that a' without a is equal in area to b' without b, and so by a further 'subtraction' that a and b are equal in area. (Alternatively we could proceed by drawing a congruent copy of b adjacent to or overlapping a and then making repeated subtractions from the whole figure containing a and the congruent copy of b.) However, it is clear that these assumptions are not by themselves sufficient to show that area is a continuous quantity, and to establish this point we must now turn to our correlation between plane figures and line-segments.

We observe first that if we arbitrarily select any line-segment AB, then all rectangles on a base equal in length to AB can be correlated in the required way, for all we need to do here is to correlate the rectangle with any line-segment equal to its height. Rectangles on the required base which are thus correlated with equal line-segments are evidently congruent, and hence by (i) equal in area. Also if R_1 is a rectangle thus correlated with a line-segment that is less than the line-segment correlated with another rectangle R_2, then evidently a congruent copy of R_1 can be drawn as a proper part of R_2, so by (i) and (ii) R_1 is less in area than R_2. Our correlation does therefore preserve the intuitive understanding of the relations 'equal in area', 'less in area', and 'greater in area'[22], and it would be easy to go on to show that it preserved the intuitive understanding of one area being the sum of two others.

[22] This is another application of the 'law of closed systems' (footnote 18, p. 216).

The next step is to extend the correlation to all rectangles, whether or not they are on a base equal to our chosen base, and for this we need to show that for any rectangle another can be found which has the same area and is of any required base. Let ABCD be any rectangle, and on AB produced mark BX

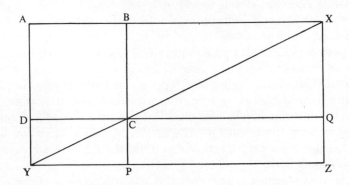

of the same length as the base required. Join XC and produce to meet AD produced at Y. Complete the rectangle YAXZ, produce BC to meet YZ at P and DC to meet XZ at Q. Then PCQZ is a rectangle on the required base, and we have to show that it is equal in area to ABCD. But this is easily seen from the observation that triangles XZY and YAX are congruent (and so by (i) equal in area), and so too are triangles CPY and YDC, and similarly triangles XQC and CBX. From this the desired result follows at once by two applications of our assumption (iii) that equals taken from equals leave equals.

So far, then, we have shown that all rectangles can be correlated in the required way with line-segments. We have only to select some particular line-segment arbitrarily, correlate each rectangle with an equal rectangle on a base of the same length as our selected line-segment, and then correlate each of the latter rectangles with a line-segment equal to its height. We can then define equality and addition for areas of rectangles in the way suggested (which, as we have seen, will coincide with our ordinary understanding) and it will follow that area *of rectangles* is a continuous quantity. It is also easy to see that the definitions of equality and addition of area are in effect independent of the particular line-segment chosen as a 'base' for the correlation, in that choice of any other line-segment would yield equivalent definitions. (At the cost of complicating the correlation we could in the present case have avoided beginning with an arbitrary line-segment[23], but the present technique is conveniently simple and we shall want to re-employ it later.) The next step is therefore to extend our results to plane figures other than rectangles. In fact the extension to other rectilinear figures presents no difficulty, for, using (iii), it is easy to show that all parallelograms, and therefore all triangles, can be correlated with rectangles of the same area and then to extend this result to polygonal figures generally by showing how each polygonal figure can be represented as a sum of triangles. What remains is to consider curvilinear figures, and for definiteness we may first consider the circle, showing that for each circle there exists a polygonal figure (and therefore a rectangle) of the same area.[24].

The first step is to show that it is possible to approximate the area of the circle as closely as we please by inscribed or circumscribed polygons, and this we do by showing how to inscribe and circumscribe polygons about the circle in such a way that for any polygonal figure z that may be specified there will exist an inscribed and a circumscribed polygon which differ in area by less than the area of z. For this lemma we may conveniently use Euclid's own construction of a series of inscribed

---

[23] We could have begun by correlating *squares* with the line-segments that are their sides, and then where S_1 and S_2 are squares we should construe $S_1 \oplus S_2$ as the square on the hypotenuse of a right-angled triangle whose other sides are correlated with S_1 and S_2. Using Pythagoras theorem we see that this gives the required result in this case, and we also see that for any squares S_1 and S_2 (with $S_1 \succ S_2$) there is always a square S_3 such that $S_1 \approx S_2 \oplus S_3$. Then as the next step we would correlate any rectangle which has sides AB and BC (with AB $\succ$ BC) with the square S_3 such that if S_1 has side $\frac{1}{2} \cdot (AB \oplus BC)$ and S_2 has side $\frac{1}{2} \cdot (AB \ominus BC)$, then $S_1 \approx S_2 \oplus S_3$. It is easily checked that the rectangle is equal in area to the correlated square.

[24] It has often been remarked that the famous proof in Euclid XII, 1–2 that circles are to one another as the squares on their diameters contains a gap, in that Euclid assumes without justification that for any squares S_1 and S_2 and any circle C there will be a plane figure X such that as S_1 is in area to S_2 so C is in area to X. The present considerations will in effect plug that gap, for when we have shown that for any circle C there is a polygonal figure of the same area as C the proposition will follow from the correlation of polygonal figures with line-segments.

polygons (Book XII, Prop. 2), adding a complementary series of circumscribed polygons, as Archimedes does[25]. So, given any circle, we first inscribe a square and add a circumscribing square by drawing tangents at the points of contact. Next on each side of the inscribed square we erect an isosceles triangle with its vertex on the circle, thus increasing the inscribed figure

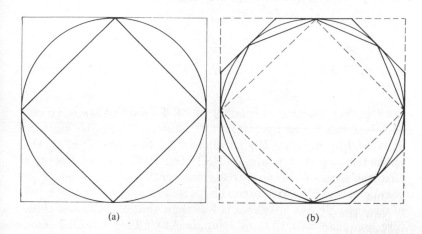

(a)   (b)

to an octagon, and at the same time add tangents at the new points of contact to form a circumscribing octagon. We continue to inscribe larger polygons and to circumscribe smaller polygons on this plan as far as desired. Now we have to show that the area of the triangles between the inscribed and circumscribed polygons is so diminished at each step of the construction that it eventually becomes less than that of any polygonal figure you may specify, and this evidently results (via Archimedes' axiom) from the observation that at each step one such triangle is replaced by another two which are together less than half the area of the original triangle. We prove that as follows. Let ABC be any triangle between two inner and outer polygons, with B and C the points of contact with the circle. Let P be the newly introduced point of contact, and X and Y the points where the tangent at P meets AB and AC respectively. We have to show that the triangles BXP and PYC are together less than half the area of triangle ABC (in fact they

[25] Archimedes, *On the measurement of the circle* (in Heath 1897).

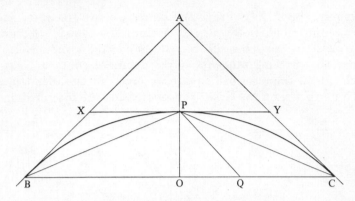

are together less than a quarter of it). To see this we have only to bisect the figure by joining AP and producing to meet BC at O, and by drawing PQ parallel to AC to meet BC at Q. It is readily shown that triangles BXP and PQC are congruent, so that the area of the two triangles together is the area of the parallelogram PYCQ. Since this parallelogram is a proper part of the triangle AOC, its area is less than that of AOC and therefore less than half that of triangle ABC. This establishes our lemma.

Now the main theorem may be proved thus. Any inscribed polygon is evidently a proper part of, and therefore less in area than, any circumscribed polygon. Therefore the areas of the inscribed polygons have an upper bound, and the areas of the circumscribed polygons a lower bound. However, we have already shown that the area of polygonal figures is a continuous quantity and therefore there will exist a polygonal figure, say x, which is (in area) a least upper bound of the inscribed polygons, and there will exist a polygonal figure, say y, which is (in area) a greatest lower bound of the circumscribed polygons. Furthermore, x and y must in fact be equal in area to one another and to the circle, for since each inscribed polygon is a proper part of the circle, and therefore less in area than the circle, the area of x must be less than or equal to that of the circle. Similarly the area of the circle is less than or equal to the area of y. Therefore what needs to be shown is that the area of y is less than or equal to that of x. However, this is an immediate consequence of the lemma, for if the area of y were greater than that of x there would be a polygonal figure, say z, with an area which was the difference between the two, and every inscribed polygon would differ from every circumscribed polygon at least by the area of z. But according to our lemma that is impossible. So this completes the proof, and we conclude that for every circle there exists a polygonal figure (and therefore a rectangle to any specified base) which has the same area. Of course it is well known that such a polygonal figure cannot be *constructed with ruler and compass*, but our assumptions—and particularly the assumption of continuity for length of line-segments—show that there must be one.

The method we have used to demonstrate this fact about the circle can evidently be applied to any *ordinary* curvilinear plane figure, since it is obvious that any *ordinary* closed curve can be approximated by inscribed and circumscribed polygons as closely as we please. However, it remains a question—and a question mainly calling for a definition of 'curve'—whether we can show that this holds for all curves. For example it presumably would not hold of the 'remarkable' curves discovered by Weierstrass and Peano[26]. However, I shall not pursue this topic any further. It may also be remarked that from a formal point of view this argument could be said to be unnecessary anyway, since we might simply *define* the area of a curvilinear figure outright as the area of any least upper bound of all the rectilinear polygons that can be inscribed in it, and leave it at that. But this basic stipulation would not be very satisfactory by itself, for it would not then be clear whether the stipulation conformed to our intuitive understanding of area—whether for example the area of a circle (so defined) would indeed be less than the area of any circumscribed polygon. When we came to prove this point we should be back, in effect, to the argument I have just been offering.

The purpose of this little excursion into the foundations of Euclidean geometry has been to illustrate the process (which we shall shortly apply with a rather different example) whereby we may argue that if one quantity is a continuous quantity then so is another. In modern axiomatizations of Euclidean geometry continuity will generally be assumed only for the length of straight line-segments, and from this basis it will be

[26] See for example Waismann 1951, Chap. 12.

proved that the same holds for length of (ordinary) curved line-segments, for the area of rectilinear and (ordinary) curvilinear plane figures and surfaces of solids, and again for the volume of polyhedra and other solids. All these quantities of Euclidean geometry will be continuous, and will also be unbounded provided that the length of line-segments is unbounded.

Now let us return to our main topic, which is the theory of proportion and its application to geometrical quantities. First I shall illustrate the definition of sameness of ratio with a proof, resting directly on the definition, of a fundamental theorem in the theory of similar triangles: if triangles ABC and XYZ have corresponding angles equal, then they also have their corresponding sides proportionate, i.e.

if $A\hat{B}C \approx X\hat{Y}Z$, $B\hat{C}A \approx Y\hat{Z}X$, and (hence) $C\hat{A}B \approx Z\hat{X}Y$,

then $AB :_{\text{length}} XY = AC :_{\text{length}} XZ = BC :_{\text{length}} YZ$.

(I assume the relevant properties of parallel lines and congruent triangles.)

As a lemma we show that if in any triangle OAB the side OA be divided into any number of equal parts OA_1, A_1A_2. A_2A_3, A_3A_4, ..., and lines parallel to the base AB be drawn

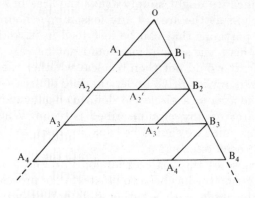

from A_1, A_2, A_3, A_4, ... to meet OB at B_1, B_2, B_3, B_4, ... respectively, then also the side OB is divided into the same number of equal parts, OB_1, B_1B_2, B_2B_3, B_3B_4, ... To prove this we construct parallel to the side OA lines from B_1, B_2, B_3 ... to meet A_2B_2, A_3B_3, A_4B_4, ... at A_2', A_3', A_4', ...

respectively. Then first we observe that each of these lines B_1A_2', B_2A_3', B_3A_4', ... is equal to the corresponding line A_1A_2, A_2A_3, A_3A_4, ..., since they are opposite sides of a parallelogram, and therefore that all these lines are equal to one another and to OA_1. Next we observe that all the triangles OA_1B_1, $B_1A_2'B_2$, $B_2A_3'B_3$, $B_3A_4'B_4$, ... have their corresponding angles equal, as is easily seen from the properties of parallel lines. It follows at once that all these triangles are congruent, and hence that OB_1, B_1B_2, B_2B_3, B_3B_4, ... are all equal, which was to be proved.

Now we proceed to the main proposition, which we may formulate thus: if OAB is any triangle and PQ a line parallel

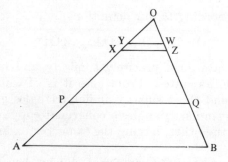

to AB meeting OA and OB at P and Q respectively, then

$$OA :_{\text{length}} OP = OB :_{\text{length}} OQ.$$

For suppose that

$$OA :_{\text{length}} OP \geqslant n : m.$$

Since length is a continuous quantity, this is to say that

$$\frac{1}{n} \circ OA \not\geqslant_{\text{length}} \frac{1}{m} \circ OP.$$

If then we mark OX on OA equal to an nth part of OA, and OY on OP equal to an mth part of OP, Y must either coincide with X or fall on OX. Now draw XZ parallel to AB to meet OB at Z, and YW parallel to PQ to meet OQ at W. Then since AB and PQ are parallel, so are XZ and YW, and hence W either coincides with Z or falls on OZ. In either case we have

$$OZ \not\geqslant_{\text{length}} OW$$

but by our lemma

$$OZ \approx_{\text{length}} \frac{1}{n} \circ OB, \quad OW \approx_{\text{length}} \frac{1}{m} \circ OQ.$$

Hence

$$\frac{1}{n} \circ OB \succcurlyeq_{\text{length}} \frac{1}{m} \circ OQ$$

or in other words

$$OB :_{\text{length}} OQ \geqslant n : m.$$

Conditionalizing this argument, we have shown that

$$(\forall nm)(OA :_{\text{length}} OP \geqslant n : m \to OB :_{\text{length}} OQ \geqslant n : m)$$

and that is, according to our definition,

$$OB :_{\text{length}} OQ \geqslant OA :_{\text{length}} OP.$$

The converse is argued precisely similarly and hence follows the proposition as stated. From here it is of course a simple matter to extend the result to similar triangles generally, for given any two similar triangles a congruent copy of one can be drawn within the other, having the same vertex and a parallel base.

The argument has been somewhat clumsy, because I wished to illustrate a direct use of the definition of sameness of ratio, and it could evidently be streamlined. For instance we might notice that in any triangle OAB lines drawn parallel to the base AB will associated with each segment of the side OA one and only one segment of the side OB. Now the proof of the lemma relied only on the fact that equal segments of OA were thus associated with equal segments of OB, and discrete segments of OA with discrete segments of OB, for from this it follows that a segment which is one n^{th} of OA is associated with a segment which is one n^{th} of OB. However, it also follows that if x and y are any segments of OA, and $f(x)$ and $f(y)$ the associated segments of OB, then $f(x \oplus y)$, the segment associated with $x \oplus y$, is $f(x) \oplus f(y)$. And from this it follows that if $x \succcurlyeq y$ then $f(x) \succcurlyeq f(y)$, and conversely, which was the point used in the main step of the proof. For suppose we grant

(i) $x \succcurlyeq y \to f(x) \succcurlyeq f(y)$

(ii) $f\left(\dfrac{1}{n}\circ x\right)\approx\dfrac{1}{n}\circ f(x)$

then the argument was that if

$$\frac{1}{n}\circ x \succcurlyeq \frac{1}{m}\circ y$$

then by (i)

$$f\left(\frac{1}{n}\circ x\right)\succcurlyeq f\left(\frac{1}{m}\circ y\right)$$

and therefore by (ii)

$$\frac{1}{n}\circ f(x)\succcurlyeq\frac{1}{m}\circ f(y)$$

so, conditionalizing and applying the definition, we have

$$f(x):f(y)\geqslant x:y.$$

The converse is similarly established by applying the converse of (i) above.

We could therefore generalize the principle of this proof as follows. Suppose we have any set α (in our example, line-segments of OA) whose members form a complete set of instances of some continuous quantity (in our example, length), i.e. the postulates for a continuous quantity (not necessarily unbounded) must be satisfied when the variables are taken as restricted to the members of α. Suppose we can also find a one–one correlation f, correlating each member x of α with a member $f(x)$ of another set β (in our example, segments of OB), and which is such that

$$x\approx y\leftrightarrow f(x)\approx(fy)$$

$$f(x\oplus y)\approx f(x)\oplus f(y).$$

Then without more ado we can conclude that

$$x:y=f(x):f(y).$$

As a further extension of this general theorem we may notice that the quantity of which the members of α are instances need not be the same as the quantity of which the members of β are instances, for although it was so in our example the proof is not affected if the quantities are taken to be different. An illustration of this extended theorem would be its application to show that rectangles on the same base are to one another in area as their heights are to one another in length, i.e. that if R_1 and R_2 are any rectangles on the same base and h_1 and h_2 are the line-segments which are their heights, then

$$R_1 :_{\text{area}} R_2 = h_1 :_{\text{length}} h_2.$$

The required correlation is of course precisely that which we used earlier when proving that rectangular area is a continuous quantity.

Such one–one correlations are often useful when operating with proportions, and as another example we may take the principle of the Greek method of exhaustion which was used to obtain results about the areas of curvilinear figures. I shall develop this by continuing the discussion of circular area (pp. 222–5), and sketching a proof that circles are to one another in area as the squares on their diameters (or radii). Suppose then that we are given two circles C_1 and C_2, and that we inscribe in each the series of regular polygons described earlier. We must first show that if S_1 and S_2 are the squares on the diameters (or radii) of each circle, $P_1{}^n$ is an n-sided polygon inscribed in C_1, and $P_2{}^n$ is the similar n-sided polygon inscribed in C_2, then

$$P_1{}^n :_{\text{area}} P_2{}^n = S_1 :_{\text{area}} S_2.$$

The proof of this is quite straightforward, but it is perhaps worth sketching for the sake of illustrating the technique of arguing with proportions.

We shall need to establish two lemmas. The first is a simple corollary to the basic thesis about rectangular areas just noted, and is that rectangles which are equal in area have their heights in inverse proportion to their bases, i.e. if R_1 and R_2 are rectangles of equal area, with heights h_1 and h_2 and bases b_1 and b_2 respectively, then

$$h_1 :_{\text{length}} h_2 = b_2 :_{\text{length}} b_1.$$

This is easily proved by considering a further rectangle Q with height h_2 and base b_1. For, comparing Q and R_1, we have

$$R_1 :_{\text{area}} Q = h_1 :_{\text{length}} h_2$$

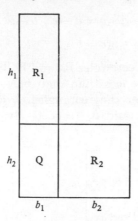

and comparing Q and R_2 we have

$$R_2 :_\text{area} Q = b_2 :_\text{length} b_1.$$

But by hypothesis

$$R_1 \approx_\text{area} R_2$$

whence

$$R_1 :_\text{area} Q = R_2 :_\text{area} Q$$

and so

$$h_1 :_\text{length} h_2 = b_2 :_\text{length} b_1.$$

We now use this result to establish our second lemma, that similar rectangles are to one another in area as the squares on their corresponding sides, i.e. if R_1 and R_2 are similar rectangles, so that

$$h_1 :_\text{length} h_2 = b_1 :_\text{length} b_2$$

and S_1 and S_2 are squares of side b_1 and b_2 respectively, then

$$R_1 :_\text{area} R_2 = S_1 :_\text{area} S_2.$$

To prove this we shall need three theorems from the general theory of proportion, which I quote here without proof[27]:

(i) $a :_\varphi b = c :_\varphi d \rightarrow a :_\varphi b = a \oplus c :_\varphi b \oplus d$

---

[27] Strictly speaking the first of these theorems requires the existential condition $E\,!a \oplus c$ & $E\,!b \oplus d$, and the second $E\,!a \ominus c$ & $E\,!b \ominus d$. The first of these is always satisfied for geometrical quantities, and the second is here assured by the stipulation that $a \succ c$ and $b \succ d$.

(ii) $a :_\varphi b = c :_\varphi d \to a :_\varphi b = a \ominus c :_\varphi b \ominus d$

(iii) $a :_\varphi b = c :_\varphi d \to a :_\varphi c = b :_\varphi d$.

Now let two similar rectangles R_1 and R_2 be given, and suppose that their heights are less than their bases. (The contrary case is argued dually.) Let the rectangles C_1 and C_2, of heights c_1 and c_2 and bases b_1 and b_2 respectively, be those needed to

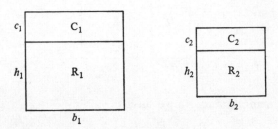

complete the squares on b_1 and b_2. Then what we have to show is (omitting subscripts for brevity)

$$R_1 : R_2 = R_1 \oplus C_1 : R_2 \oplus C_2.$$

First we notice that by the hypothesis of similarity, and using (ii) above, we have

$$h_1 : h_2 = b_1 : b_2 = b_1 \ominus h_1 : b_2 \ominus h_2$$

and by construction we have

$$c_1 \approx b_1 \ominus h_1, \quad c_2 \approx b_2 \ominus h_2$$

whence

$$h_1 : h_2 = c_1 : c_2$$

or, using (iii),

$$h_1 : c_1 = h_2 : c_2.$$

Next we notice that R_1 and C_1 are rectangles on the same base b_1, and equally R_2 and C_2 are rectangles on the same base b_2, and so we have

$$R_1 : C_1 = h_1 : c_1, \quad R_2 : C_2 = h_2 : c_2.$$

Putting this point together with the last we deduce

$$R_1 : C_1 = R_2 : C_2$$

or, using (iii),

$$R_1:R_2 = C_1:C_2$$

and therefore using (i)

$$R_1:R_2 = R_1 \oplus C_1:R_2 \oplus C_2$$

which was to be proved.

Now from this lemma about similar rectangles we can easily show that the same holds for similar triangles, i.e. that similar triangles are to one another in area as the squares on their corresponding sides, for the area of a triangle is half that of a rectangle on the same base and with the same height. Also, in view of the fact that every polygon can be triangulated, the result may then be extended to similar polygons generally. The form in which this result is required for our theorem is rather that if S_1 and S_2 are the squares on the diameters (or radii) of circles C_1 and C_2, and if $P_1{}^i$ and $P_2{}^i$ are any similar polygons inscribed in C_1 and C_2 respectively, then

$$P_1{}^i:P_2{}^i = S_1:S_2.$$

However, it is now clear that this can be shown without difficulty from our lemmas, so we may proceed to the main theorem.

For brevity let us use $P_n{}^1$, $P_n{}^2$, ..., $P_n{}^i$, ... for the series of polygons inscribed in the circle C_n, and $lub\,\{P_n{}^i\}$ for a polygon which is (in area) a least upper bound of this whole series. The result established in our earlier argument (pp. 222–5) is that

$$C_1 \approx lub\,\{P_1{}^i\}, \quad C_2 \approx lub\,\{P_2{}^i\}$$

and we have so far sketched a proof that

$$(\forall i)(P_1{}^i:P_2{}^i = S_1:S_2).$$

What is needed to complete the argument is therefore to show that this in turn implies

$$lub\,\{P_1{}^i\}:lub\,\{P_2{}^i\} = S_1:S_2$$

and it is here that we make use again of one–one correlations. For if we now let f be the correlation that associates with each polygon $P_1{}^i$ the polygon $f(P_1{}^i) = P_2{}^i$, the desired result will be a special case of the more general theorem:

if $(\forall i)(x^i : f(x^i) = a : b)$

then $lub\{x^i\} : lub\{f(x^i)\} = a : b.$

This general theorem is proved quite straightforwardly. Suppose

$$lub\{x^i\} : lub\{f(x^i)\} > n : m$$

and therefore (assuming the quantity concerned is continuous)

$$\frac{1}{n} \circ lub\{x^i\} \succ \frac{1}{m} \circ lub\{f(x^i)\}.$$

Then it is easily seen to follow that

$$(\exists i)(\frac{1}{n} \circ x^i \succ \frac{1}{m} \circ lub\{f(x^i)\})$$

and from this in turn that

$$(\exists i)(\frac{1}{n} \circ x^i \succ \frac{1}{m} \circ f(x^i))$$

i.e. that

$$(\exists i)(x^i : f(x^i) > n : m).$$

However, by hypothesis

$$(\forall i)(x^i : f(x^i) = a : b)$$

and therefore

$$a : b > n : m.$$

Conditionalizing this argument we have shown that

$$(\forall nm)(lub\{x^i\} : lub\{f(x^i)\} > n : m \rightarrow a : b > n : m)$$

whence it follows that

$$a : b \geqslant lub\{x^i\} : lub\{f(x^i)\}.$$

The converse is argued similarly, and this establishes our theorem—a theorem which is, as I say, the main principle of the Greek method of exhaustion.

Now let us pause to review the progress of the discussion so far. It is common practice nowadays to presuppose a knowledge of the real-number system when entering upon the study of geometry and to assign real numbers to line-segments and plane figures as measures of their length and area. The theorems that we have been developing about proportions would then be stated instead as theorems concerning the relations of the real numbers thus assigned. For example our theorem concerning similar triangles might nowadays be put in the form:

If triangles OAB and OPQ are as stated (p. 227), and if x, y, z, and w are the real numbers assigned as measures of length to OA, OP, OB, and OQ respectively, then

$$\frac{x}{y} = \frac{z}{w}, \quad \text{i.e. } x \cdot w = z \cdot y.$$

Similarly our theorem about circular area might be put in the form

There is a real number, call it π, such that for *any* circle, if x and y are the real numbers assigned as measures of area to the circle and to the square on its radius, then

$$\frac{x}{y} = \pi, \quad \text{i.e. } x = \pi \cdot y.$$

More usually we should rely on the fact that the real number assigned to a square as a measure of its area is the square of the real number assigned to its side as a measure of its length, and therefore say that if x is the measure of the area of the circle and r the measure of the length of its radius, then

$$x = \pi \cdot r^2.$$

Now our object in this discussion was to obtain a clear view of how real numbers are required in geometry, and to this end we have been seeing how we can get on without them, using instead the theory of proportion first developed by the Greek geometers. So the question arises: Is there any essential loss or gain when our theorems are put now in one way and now in the other? So far as the first theorem is concerned, no gain or loss is apparent, and it seems clear that the theory of similar triangles can be developed just as well in the theory of proportion as it can with the help of real numbers. However, it could be argued that there is a loss of information when we state the second theorem without using real numbers.

This loss does not concern the use of the real number π, for we can certainly replace the use of a constant real number by the use of a constant ratio. Our original way of stating the theorem about circular area evidently yields (via (iii) above) the result that if C_1 and C_2 are any two circles, and S_1 and S_2 are the squares on their radii, then

$$C_1 :_{\text{area}} S_1 = C_2 :_{\text{area}} S_2$$

which is to say that the ratio between the circle and the square on its radius is always the same. Therefore we may introduce a name for this ratio, say π again, and write our result as follows: for any circle C and the square on its radius S

$$C :_{\text{area}} S = \pi.$$

We may then go on to prove that the same ratio also obtains between, say, the length of the circumference of a circle and its diameter, for it is perfectly possible to show, using the theory of proportion, that[28]

$$C :_{\text{area}} S = \text{circumference of } C :_{\text{length}} \text{ diameter of } C$$

The same holds for other relationships into which π enters. Nor is there any difficulty of principle in obtaining a numerical value for π. For example it is easy to see that the ratio π must be less than the ratio of our circumscribed square to the square on the radius, and greater than the ratio of our inscribed square to the square on the radius. Hence

$$4 : 1 > \pi > 2 : 1.$$

By calculating (approximately[29]) the areas of the circumscribed and inscribed octagons we can improve this result to

$$33 : 10 > \pi > 28 : 10.$$

We may obtain closer and closer approximations by continuing

---

[28] See Archimedes, *On the measurement of the circle* (in Heath 1897).

[29] This calculation uses the ancient approximation to $\sqrt{2}$

$$\left(\frac{99}{70}\right)^2 > 2 > \left(\frac{98}{70}\right)^2$$

or, to be more exact, it uses the fact that for any square S

$$99 : 70 > \text{diagonal of } S :_{\text{length}} \text{ side of } S > 98 : 70.$$

(More convenient ways of approximating π are certainly available, as for example in Archimedes, *op. cit.*)

in this way through our series of circumscribed and inscribed polygons. Generally, then, we can replace the geometrical theses in which π is used to denote a real number by theses in which it is used to denote a ratio, and no loss need be expected in this direction.

What we are not yet able to reproduce in the theory of proportion is the relation between lengths and areas of which a special case is the principle that if r is the measure of the length of the side of a square, then r^2 is the measure of its area, and conversely. The more general relation is (as we usually say) that the area of a rectangle is 'the base multiplied by the height', but though it makes perfectly good sense to multiply two real numbers together, the product being again a real number, there is no evident sense to the suggestion that two *lines* may be 'multiplied together' to yield a plane figure as result. What we can do, however, is to multiply together not lines, or plane figures, but *ratios*. So let us turn now to a further development of the general theory of ratios.

When proving that rectangular area was a continuous quantity, we proceeded by first selecting some line-segment arbitrarily and then using this as a base to set up a correlation between rectangles and line-segments in such a way that rectangles of greater, equal, or lesser area were correlated with line-segments of greater, equal, or lesser length. This technique suggests a method of showing that ratios themselves may be conceived as instances of a quantity, say magnitude, which also is continuous. We select any one continuous quantity, say length, and again select some particular instance arbitrarily, and then we use this to set up a correlation between ratios and line-segments in such a way that greater, equal, or lesser ratios are correlated with line-segments that are greater, equal, or lesser in length. The fundamental theorem that we rely on is that if b is any selected line-segment, and ϕ-ness any desired quantity with instances c, d, then

$$(\exists x)(c :_\varphi d = x :_\text{length} b).$$

This theorem, which affirms the existence of a 'fourth proportional' for any three terms b, c, d, is crucial to the theory of proportion but notoriously missing from Euclid's treatment. I discuss its significance further in the Appendix to this Chapter.

As stated here, it relies on the hypothesis that length is a continuous quantity, and an unbounded one, and we may continue to assume this for the time being[30].

The proof of the theorem is entirely analogous to our earlier proof that any instance of a continuous quantity can be divided into any desired proper fraction, however minute (pp. 173–7). We consider those line-segments x such that

$$c :_\varphi d \geqslant x :_\text{length} b.$$

It is easy to show that there is at least one line-segment x which satisfies this condition (assuming that length is a continuous quantity) and that there is at least one line-segment x which does not (assuming that length is an unbounded quantity). Hence the line-segments which do satisfy the condition form a non-empty and bounded set of line-segments and will therefore have a least upper bound, say a, such that

$$(\forall y)(a \leqslant_\text{length} y \leftrightarrow (\forall x)(c :_\varphi d \geqslant x :_\text{length} b \rightarrow x \leqslant_\text{length} y)).$$

It easily follows that

$$c :_\varphi d = a :_\text{length} b$$

which was to be proved.

The theorem is of some independent interest because of its bearing on the adequacy of our criterion for a continuous quantity, *viz.* that the quantity should satisfy our axioms (including L.U.B.) and have no least instance. A simple reformulation of the argument just sketched will enable us to dispense with the hypothesis that length is an unbounded quantity, and yield a proof depending only on the hypothesis of continuity for length, that for any line-segment x

$$(\exists yz)(x \approx_\text{length} y \oplus z \,\&\, c :_\varphi d = y :_\text{length} z)$$

Therefore, generalizing from length to any quantity, it follows that any instance of any quantity that is continuous according to our criterion can be divided in any desired ratio, provided there is some instance $c :_\varphi d$ of that ratio somewhere. I think that it is an important aspect of our intuitive grasp of continuity that any instance of a continuous quantity can be

---

[30] The assumption will be dropped in the next section, and we shall prove a theorem that takes its place (pp. 249–53).

divided in any desired ratio—rational or irrational—and that is now established for *all* quantities that are continuous according to our criterion, provided that there is at least *one* quantity φ-ness which exemplifies all possible ratios. That there is at least one such quantity will be argued later (pp. 248–53).

However, this is somewhat of a digression from our current purpose, which is to show how the argument establishing that rectangular area is a continuous quantity can be reapplied to show that magnitude of ratios is also a continuous quantity. Using the theorem

$$(\exists x)(c :_\varphi d = x :_{\text{length}} b)$$

we see that any ratio $x :_\varphi y$ may be correlated with the line-segments a such that

$$x :_\varphi y = a :_{\text{length}} b$$

and every ratio will thus be correlated with some line-segment, and every line-segment a will be correlated with some ratio (*viz.* the ratio $a :_{\text{length}} b$). Further, we have as a general theorem

$$a :_\varphi b \geqslant c :_\varphi b \leftrightarrow a \succcurlyeq_\varphi c$$

and this shows that the correlation satisfies our further condition that

$$(x :_\varphi y = a :_{\text{length}} b \ \& \ z :_\psi w = c :_{\text{length}} b)$$
$$\rightarrow (x :_\varphi y \geqslant z :_\psi w \leftrightarrow a \succcurlyeq_{\text{length}} c).$$

Accordingly we may now go on to define the *addition* of two ratios as before, by means of the addition of their correlated line-segments, and then the properties of addition of line-segments will carry over automatically to addition of ratios, and we have our proof that the quantity thus introduced, *viz.* magnitude of ratios, is itself a continuous quantity.

In fact it turns out once more that this definition of addition of ratios is in effect independent of the particular line-segment chosen as base of the correlation, and is independent of the selected quantity length. For suppose one correlation, using the quantity φ-ness and some fixed instance b, matches $x:y$ with a and $z:w$ with c, so that on that correlation $(x:y) + (z:w)$ is the ratio correlated with $a \oplus c$, i.e. the ratio $a \oplus c :_\varphi b$. If we now consider a different correlation, using a quantity ψ-ness and a fixed instance b', and on that correlation $x:y$ is matched with a' and $z:w$ with c', then on that correlation $(x:y) + (z:w)$ would be the ratio $a' \oplus c' :_\psi b'$. However, these must be the same ratio. For from the fact that $x:y$ is correlated both with a and with a', and similarly $z:w$ both with c and with c', we have

$$a :_\varphi b = a' :_\psi b, \quad c :_\varphi b = c' :_\psi b'.$$

Also from the assumption that ϕ-ness and ψ-ness are both unbounded quantities we have

$$\mathrm{E}!a \oplus c, \quad \mathrm{E}!a' \oplus c'$$

and these two premises together entail

$$a \oplus c :_\varphi b = a' \oplus c' :_\psi b'.$$

This result evidently suggests that it would be more elegant and equally satisfactory to offer a formal definition of addition that did not invoke any particular correlation, so we shall in fact define

$$(a:b) + (c:d) = u:v$$

(where the omitted subscripts may be the same or different) as short for

$$(\exists\phi)(\exists xyz)(\mathrm{E}!x \oplus y \ \& \ a:b = x:_\varphi z \ \& \ c:d = y:_\varphi z$$
$$\& \ u:v = x \oplus y :_\varphi z).$$

In view of the theorem just noted it follows that the ratio $u:v$ is unique, and we can also show that it always exists, i.e. that there is a quantity ϕ-ness with instances x, y, z such that

$$\mathrm{E}!x \oplus y \ \& \ a:b = x:_\varphi z \ \& \ c:d = y:_\varphi z.$$

Clearly, if there is a continuous quantity, we can be sure that it has instances x, y, z satisfying this thesis, even if it is not unbounded, simply by choosing z sufficiently small. On the other hand if there is no continuous quantity then every quantity is discrete, and in that case the thesis will be satisfied just by taking x, y, z as natural numbers. Generalizing this line of argument, it can be seen that the desired properties of addition of ratios will follow from this definition whether or not there are any continuous quantities, but we cannot afford to remain uncommitted on this point. For to show that the ratios do form an extensive quantity in our sense we also need to show that they satisfy the principle of least upper bounds, and that they would not do if all quantities were discrete. For the moment, however, let us leave that point on one side and return to our further development of the theory of ratios.

With addition of ratios now introduced, we can go on to consider ratios between the ratios themselves, which are of course defined in the same way as for any other quantity. Thus we may consider formulae such as[31]

$$a :_\varphi b = (u :_\psi v) : (c :_\chi d)$$

and we may for example prove that

$$a :_\varphi b = (a :_\varphi c) : (b :_\varphi c).$$

Until now we have had no significant formulae of the pattern $x = y : z$, but now that such formulae are available the evident analogy between the sign ': ' and the sign for division of numbers provides a powerful motive for introducing for ratios an inverse operation to correspond to multiplication of numbers. Accordingly we may allow ourselves to rewrite the last theorem as

$$(a :_\varphi b) \cdot (b :_\varphi c) = a :_\varphi c.$$

More generally, we may introduce

$$(a : b) \cdot (c : d) = u : v$$

(where the omitted subscripts may be the same or different) as another way of writing

$$a : b = (u : v) : (c : d)$$

or, what comes to the same thing, we may proceed by analogy with our definition of addition and give as definiens

$$(\exists \phi)(\exists xyz)(a : b = x :_\varphi y \ \& \ c : d = y :_\varphi z \ \& \ u : v = x :_\varphi z).$$

In old-fashioned terminology the ratio $(a : b) \cdot (c : d)$ is said to be the ratio *compounded from* the ratios $a : b$ and $c : d$, and it is

---

[31] The relevant quantity to be subscripted to the bare occurrence of : in this formula is of course magnitude of ratios. I shall always omit this subscript, just as I always omit the subscript for magnitude of natural numbers in the expression $n : m$.

not hard to show that this operation of composition has the expected formal properties of multiplication.

Now let us return to the statement 'the area of a rectangle is the base multiplied by the height'. The analogue of this statement in the theory of proportion is now that rectangles are to one another in area in the ratio compounded of the ratios of their heights and their bases, i.e. if R_1 and R_2 are rectangles, with heights h_1 and h_2 and bases b_1 and b_2 respectively, then

$$R_1 :_{\text{area}} R_2 = (h_1 :_{\text{length}} h_2) \cdot (b_1 :_{\text{length}} b_2).$$

The proof may be given very simply, by again considering a rectangle Q of height h_2 and base b_1. Comparing R_1 and Q we have

$$R_1 :_{\text{area}} Q = h_1 :_{\text{length}} h_2$$

and comparing Q and R_2 we have

$$Q :_{\text{area}} R_2 = b_1 :_{\text{length}} b_2.$$

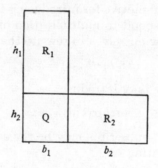

From this it is easily proved that

$$(R_1 :_{\text{area}} Q) \cdot (Q :_{\text{area}} R_2) = (h_1 :_{\text{length}} h_2) \cdot (b_1 :_{\text{length}} b_2).$$

but we have earlier said that

$$(R_1 :_{\text{area}} Q) \cdot (Q :_{\text{area}} R_2) = R_1 :_{\text{area}} R_2$$

whence follows the result. More generally, now that ratios can be added and multiplied just as real numbers can, no further difficulty need be expected in translating into the language of ratios any geometrical theses which we would nowadays state with the help of real numbers.

Indeed we can if we wish assign particular ratios to lines and figures as measures of their lengths and areas. For this purpose we have to select some standard line-segment, say l, as our 'unit' of length, and we then assign to each line-segment h the ratio $(h:_{\text{length}}l)$ as the measure of its length. Similarly we select a square S of side l as our 'unit' of area, and we assign to each plane figure P the ratio $(P:_{\text{area}}S)$ as a measure of its area. Now as a special case of the result just proved about rectangular area we evidently have

$$R:_{\text{area}}S = (h:_{\text{length}}l) \cdot (b:_{\text{length}}l)$$

and this theorem now states that if x be the ratio assigned to a rectangle as a measure of its area and if y, z be the ratios assigned to its height and base as measures of their length, then

$$x = y \cdot z.$$

The use of real numbers as measures of lengths and areas evidently presupposes the choice of some appropriate 'unit', and with the same presupposition we may use ratios to just the same purpose. Indeed by now it may seem that the real numbers and the ratios are hardly to be distinguished from one another, and that is a topic which I shall take up in the following section.

To end this section I should just like to add a coda on the use of ratios of ratios to introduce numerical comparisons which are not *fundamental* numerical comparisons, for example in respect of density or speed. The somewhat cryptic formula

$$\text{density} = \frac{\text{mass}}{\text{volume}}$$

would normally be interpreted as a rule for assigning real numbers to objects as measures of their density in terms of the real numbers assigned as measures of their masses and their volumes. However, we can equally regard it as introducing a new kind of *ratio*, *viz.* in respect of density, in terms of ratios in respect of mass and volume, in the form

$$x:_{\text{density}}y = (x:_{\text{mass}}y):(x:_{\text{volume}}y).$$

The similarly cryptic formula

$$\text{speed} = \frac{\text{distance}}{\text{time}}$$

may equally be regarded as introducing a ratio between certain processes, namely movements, to the effect that if x and y are both movements then

$$x :_{\text{speed}} y = (x :_{\text{distance}} y) : (x :_{\text{time}} y)$$

where of course the ratio $(x :_{\text{distance}} y)$ is taken as the ratio in distance (length) between the line marked out by x and the line marked out by y, and the ratio $(x :_{\text{time}} y)$ is taken as the ratio in duration between the period occupied by x and the period occupied by y[32]. (Alternatively we could, with suitable adjustments, take the subjects of these ratios to be not the movements but the moving bodies.) The advantage of this way of viewing these formulae is just that it enables us to see how we can make numerical comparisons in respect of speed or density which do not depend for their sense upon any choice of unit. We may for example say that x is twice as dense as y when

$$x :_{\text{density}} y = 2 : 1$$

i.e. when

$$(x :_{\text{mass}} y) : (x :_{\text{volume}} y) = 2 : 1$$

for fundamental numerical comparisons are available in respect of mass and volume, and in terms of these the above formula can be defined, as we have seen.

3. Real Numbers and the Theory of Proportion

The main topic of this section is to take up the question recently raised, whether the real numbers can simply be identified with the ratios we were using in the last section. However, as a preliminary it will be convenient to begin with some more general remarks on the theory of these ratios.

The postulates for the theory are the traditional postulates for extensive quantities, except that we do not include the assumption of unboundedness

$$\mathrm{E}!a \oplus b.$$

---

[32] This explanation of course concerns *average* speed. By adding suitable qualifications about direction we may obtain an account of average velocity. Instantaneous velocity may then be introduced in terms of average velocity.

The postulates we do include require first that $\approx$ be an equivalence relation, and next that $\oplus$ satisfy the theses

(1') $\mathrm{E}!\zeta \oplus \theta \rightarrow \mathrm{E}!\zeta$

(2') $\mathrm{E}!a \oplus b \rightarrow a \oplus b \approx b \oplus a$

(3') $\mathrm{E}!a \oplus (b \oplus c) \rightarrow a \oplus (b \oplus c) \approx (a \oplus b) \oplus c$

(4') $\mathrm{E}!a \oplus c \rightarrow (a \approx b \rightarrow a \oplus c \approx b \oplus c)$

(5a) $a \not\approx a \oplus b$

(5b) $a \not\approx b \rightarrow (\exists x)(a \approx b \oplus x \lor a \oplus x \approx b)$.

In view of (5a) and 5(b) we then define

$$a \succ b \quad \text{for} \quad (\exists x)(a \approx b \oplus x)$$

and we show that $\succ$ is a comparative relation with $\approx$ as its associated equality relation. The final postulate is then

L.U.B. $\quad (\exists x)(Fx)\ \&\ (\exists y)(\forall x)(Fx \rightarrow x \leqslant y)$

$$\rightarrow (\exists z)(\forall y)(z \leqslant y \leftrightarrow (\forall x)(Fx \rightarrow x \leqslant y)).$$

Now there are four main types of models for these postulates. One major distinction is that between models for *continuous* quantities and models for *discrete* quantities, which is the distinction between quantities that do and do not also satisfy the thesis

$$(\exists x)(a \succ x)$$

or, what comes to the same thing, the thesis

$$\mathrm{E}!\frac{1}{n}\circ a.$$

The other major distinction is that between models for *unbounded* and for *bounded* quantities, which is the distinction between quantities that do and do not also satisfy the thesis

$$\mathrm{E}!a \oplus b$$

or, what comes to the same thing, the thesis

$$\mathrm{E}!n \circ a.$$

These distinctions yield our four main types, though of course there are also some subdistinctions to be drawn.

It is easily enough seen that all models for discrete and unbounded quantities are quasi-isomorphic to one another and to the natural numbers under equality and addition, where by a *quasi*-isomorphism I mean roughly one which does not take any account of the number of distinct elements of the model which all bear (the interpretation of) the relation $\approx$ to one another. Let us say that a model is *minimal* if $a \approx b$ is interpreted as true in that model only when a and b are interpreted as the *same* element of the model. Then evidently every model can be contracted to a minimal model by selecting just one representative from each of the equivalence classes generated by the interpretation of $\approx$ in that model, and the difference between two models which may be contracted in this way to the same minimal model is of no interest. So we may say more simply that every *minimal* model for a discrete and unbounded quantity is in the ordinary sense isomorphic to the natural numbers under equality and addition. To see this we have only to observe that in the discrete case we can prove that there exists some unit element such that all other elements are integral multiples of it, i.e. we can prove that for some a

$$(\forall x)(\exists n)(x \approx n \circ a)$$

and where the quantity is unbounded we also have

$$(\forall n)(\exists x)(x \approx n \circ a).$$

The required isomorphism is thus obtained by correlating each element of the model with the natural number associated with it by these theses, and it is easy to show that this correlation has the desired properties.

The same treatment is available for *bounded* and discrete quantities, where of course we find by the same means that every minimal model is isomorphic to some finite initial segment of the natural numbers under equality and addition. The only difference between two such minimal models is therefore that one represents a quantity which has a greater or lesser number of exemplified degrees than another, and all these models may reasonably be classed together as of the same type.

Turning now to consider the continuous case we may equally

show that all minimal models for continuous and unbounded quantities are isomorphic to one another and to the positive real numbers under equality and addition. The crucial theorem here is the theorem of the existence of the fourth proportional (p. 237–8), namely that where ψ-ness is a continuous and unbounded quantity

$$\phi\alpha \ \& \ \psi\beta \rightarrow (\forall x)(\phi x \rightarrow (\exists y)(x:_\varphi a = y:_\psi b)).$$

Given any two minimal models for our postulates we may therefore set up a correlation by first arbitrarily selecting any element a of one and any element b of the other and then correlating each element x of the first model with the element y of the second model such that $x:_\varphi a = y:_\psi b$. The above theorem then assures us that so long as the quantities concerned are continuous and unbounded every element of either model is thereby correlated with some element of the other, and it is quite straightforward to show that this correlation has the desired properties. The further claim about real numbers follows of course from the fact that the real numbers under equality and addition themselves form a minimal model for a continuous and unbounded quantity.

The fourth type, namely models for bounded and continuous quantities, subdivides into two according as there is or is not a greatest element in the model, i.e. according as the thesis

$$(\exists x)(\forall y)(y \leqslant x)$$

is or is not satisfied. Where this thesis is satisfied all the minimal models are isomorphic to one another and to every initial segment of the positive real numbers up to and including some specified number, e.g. to the positive real numbers less than or equal to 1; where the thesis is not satisfied all the minimal models are isomorphic to one another and to every initial segment of the positive real numbers up to but *not* including some specified real number, e.g. to the positive real numbers less than 1. To set up the required correlation in these cases we consider the condition

$$E!x \oplus x.$$

Since the quantity concerned is continuous there is at least one element x which satisfies the condition, and since the quantity is bounded there is at least one which does not. Further, every element which satisfies it is less than every element which does not, so there is a least upper bound of all the elements satisfying the condition. Given any two minimal models, therefore, there will be one element a of the first which is the least upper bound of the elements x in that model satisfying $\mathrm{E}!x \oplus x$, and one corresponding element b of the other. We then use a and b as the 'bases' of the correlation as before, by again correlating each element x of the first model with the element y of the second such that $x:a = y:b$, and it is again easy to show that this correlation has the required properties. It is perhaps worth noting that in this case, as in the discrete cases, the isomorphism is unique, though in the continuous and unbounded case there are any number of isomorphisms depending on the initial arbitrary selection of elements a and b to be the 'bases' of the correlation.

The last of these types of model, i.e. a model for a continuous quantity that is bounded but has no greatest element, is perhaps a little unexpected. At any rate it does not seem at all plausible that any quantity capable of fundamental numerical comparison should be of this type, and there might be something to be said for ruling out such models, for example simply by laying down the extra axiom (to apply in the continuous case only)

$$(\forall x)(x \prec a \rightarrow \mathrm{E}!x \oplus x) \rightarrow \mathrm{E}!a \oplus a.$$

However, I do not think that the axiom would have any useful consequences, so it seems better to tolerate the unexpected type of model.

Now it follows from what we have just said that there are models of each of the types we have distinguished, namely the models with natural numbers as elements in the discrete cases and real numbers as elements in the continuous cases. If it is objected that it is hardly fair to make use of the real numbers at this stage, since our theory of proportion is designed as an *alternative* to the classical theory of real numbers, then we may take instead any of the representatives of real numbers used in the classical constructions, i.e. sets or sequences of rational numbers with appropriate relations and operations defined for them. However, it may then further be objected that we have as yet no right to assume an understanding of arbitrary sets or sequences of rational numbers, or indeed of arbitrary predicates of rational numbers. At any rate we must admit that the construction of rational numbers given in the last chapter permitted us to handle rational numbers only in certain special contexts, *viz.* those built up by means of $=$, $>$, $+$, $\cdot$, and a few other symbols; we have not yet introduced any variable which may be taken to range over arbitrary predicates of rational numbers.

To remedy this defect we may extend the original construction of numerical quantifiers given in the first part of this work. In that construction we made use of variables and schematic letters for predicates of pure quantifiers, taking $\Re$ as a schematic letter for dyadic predicates of this type. Now for some such dyadic predicates $\Re$ the thesis

$$\frac{n}{m}=\frac{p}{q} \to (\exists n\Re\exists m \leftrightarrow \exists p\Re\exists q)$$

will hold for all numerical quantifiers $(\exists n\alpha)(-\alpha-)$, $(\exists m\alpha)$ $(-\alpha-)$, $(\exists p\alpha)(-\alpha-)$, and $(\exists q\alpha)(-\alpha-)$. In such a case we may perfectly well regard $\Re$ instead as a monadic predicate of rational numbers, and rewrite $\exists n\Re\exists m$ as, say, $\Phi(n/m)$. From the formal point of view we shall therefore regard a formula involving the letter Φ in this use, say

$$(\forall\Phi)(-\Phi(n/m)-)$$

simply as an abbreviation for

$$(\forall\Re)\left((\forall nmpq)\left(\frac{n}{m}=\frac{p}{q} \to (\exists n\Re\exists m \leftrightarrow \exists p\Re\exists q)\right) \to (-\exists n\Re\exists m-)\right).$$

It will also simplify our formulae to introduce r, s, t, ... once more to take the place of n/m, p/q, k/l, ... as suggested on p. 79, so Φ may also occur simply in the context Φr.

There is some choice of how to construct, from predicates of rational numbers, a quantity which is continuous and unbounded. Perhaps our simplest plan will be to proceed, as in §1 (pp. 204–9), by considering the predicates Φ of rational numbers which are non-empty and bounded above, i.e. which satisfy the condition

$$(\exists r)(\Phi r) \ \& \ (\exists s)(\forall r)(\Phi r \to r \leqslant s).$$

Let us abbreviate this condition to

$$Bounded(\Phi).$$

Now we define an ordering relation on predicates by saying that Φ is to be less than or equal to Ψ when every upper bound of Ψ is also an upper bound of Φ, i.e.

$$\Phi \leqslant \Psi \quad \text{for} \quad (\forall r)((\forall s)(\Psi s \to s \leqslant r) \to (\forall s)(\Phi s \to s \leqslant r)).$$

It is then perfectly straightforward to show that this relation *is* an appropriate ordering relation, i.e. that

$$\Phi \leqslant \Psi \ \& \ \Psi \leqslant X \to \Phi \leqslant X$$

$$\Phi \leqslant \Psi \lor \Psi \leqslant \Phi.$$

In terms of this relation we may therefore define, by the standard method, a comparative relation proper and an associated equality relation

$$\Phi \approx \Psi \quad \text{for} \quad \Phi \leqslant \Psi \ \& \ \Psi \leqslant \Phi$$

$$\Phi \prec \Psi \quad \text{for} \quad \Phi \leqslant \Psi \ \& \ \Psi \nleqslant \Phi.$$

We now add a definition of an operation $\oplus$ on these predicates by putting

$$(\Phi \oplus \Psi)r \quad \text{for} \quad (\exists st)(r = s + t \ \& \ \Phi s \ \& \ \Psi t)$$

and we prove

(1) $Bounded(\Phi) \ \& \ Bounded(\Psi) \to Bounded(\Phi \oplus \Psi)$

(2) $\Phi \oplus \Psi \approx \Psi \oplus \Phi$

(3) $\Phi \oplus (\Psi \oplus X) \approx (\Phi \oplus \Psi) \oplus X$

(4) $\Phi \approx \Psi \to (\Phi \oplus X \approx \Psi \oplus X)$

(5) $Bounded(\Phi) \ \& \ Bounded(\Psi)$

$$\to (\Phi \prec \Psi \leftrightarrow (\exists X)(Bounded(X) \ \& \ \Phi \oplus X \approx \Psi)).$$

All the proofs are perfectly straightforward, and only the last presents any noticeable difficulty[33]. What remains is to show

---

[33] To establish the implication from left to right we may consider the predicate X so defined that

$$(\forall r)(Xr \leftrightarrow (\forall s)(\Phi s \to (\exists t)(\Psi t \ \& \ r + s \leqslant t))).$$

that L.U.B. is also satisfied for these bounded predicates of rational numbers, and supposing that this can be done we may then infer that they form a model for our postulates. Further, in view of (1) above the quantity in question is unbounded, and in view of the simple theorem

$$(\exists \Psi')(\Psi' \prec \Phi)$$

the quantity is continuous. We shall then have proved that there is an example of a continuous and unbounded quantity, from which it evidently follows that there are also examples of both sorts of continuous and bounded quantity.

When we turn to the proof of L.U.B. it becomes clear that in order to *state* this postulate more apparatus will be required, for we must introduce a variable which ranges over predicates applicable to these predicates of rational numbers, i.e. predicates applicable to certain dyadic relations between pure quantifiers, and no such variable has actually figured in any of our previous constructions. However, no new principle is involved, for what is wanted is simply a variable to represent the *remainders* to predicates of rational numbers, and we have already given a perfectly general discussion of remainders[34]. We shall therefore introduce new elementary schemata, say $(\mathfrak{Q}r)(\Phi r)$ and its alphabetic variants, to represent any proposition which contains a predicate Φ of rational numbers as component, remembering that this must always occur in a context Φr. The new symbol $\mathfrak{Q}$ will of course be subject to the perfectly general rules of first-order quantification theory, and in addition to the rule of substitution appropriate for remainders. (The schema $(\mathfrak{Q}r)(\Phi r)$ is of course short for $(\mathfrak{Q}nm)(\Phi(n/m))$, and, by virtue of our understanding of Φ in this use, this is a special case of the more general schema $(\mathfrak{Q}q_1 q_2)(q_1 \mathfrak{R} q_2)$ for any remainder to a dyadic relation between pure quantifiers. Therefore the considerations of Volume 1, pp. 136–41 are evidently relevant.)

The thesis that we have to prove may therefore be stated as

$$(\exists \Phi)(\mathfrak{Q}r)(\Phi r) \ \& \ (\exists \Psi')(\forall \Phi)((\mathfrak{Q}r)(\Phi r) \ \rightarrow \ \Phi \leqslant \Psi')$$

$$\rightarrow (\exists X)(\forall \Psi')(X \leqslant \Psi' \leftrightarrow (\forall \Phi)((\mathfrak{Q}r)(\Phi r) \rightarrow \Phi \leqslant \Psi'))$$

[34] Volume 1, pp. 69–72, 82–3, 94–7.

where all the quantifiers are to be understood as restricted to predicates Φ of rational numbers such that $Bounded(\Phi)$. The proof is then entirely analogous to that sketched earlier in §1 (pp. 206–8). We assume the two antecedents, which are in full

   (i) $(\exists\Phi)(Bounded(\Phi)\ \&\ (\mho r)(\Phi r))$

   (ii) $(\exists\Psi)(Bounded(\Psi)\ \&\ (\forall\Phi)(Bounded(\Phi)\ \&\ (\mho r)(\Phi r)$
$$\rightarrow \Phi \leqslant \Psi))$$

and we consider the predicate X of rational numbers such that Xs holds if and only if

$$(\exists\Phi)(Bounded(\Phi)\ \&\ (\mho r)(\Phi r)\ \&\ \Phi s).$$

It is obvious that this predicate is, in our stipulated sense, a predicate of rational numbers, and it is easily shown to be non-empty and bounded above, i.e. it is easily shown that

  (iii) $(\exists s)(\exists\Phi)(Bounded(\Phi)\ \&\ (\mho r)(\Phi r)\ \&\ \Phi s)$

  (iv) $(\exists t)(\forall s)((\exists\Phi)(Bounded(\Phi)\ \&\ (\mho r)(\Phi r)\ \&\ \Phi s)\rightarrow s \leqslant t)$

((iii) follows at once from (i) together with the first clause of the definition of $Bounded(\Phi)$, and (iv) follows from (ii) together with the second clause of that definition). However, it is also easy to see that this predicate X is the least upper bound we desire, for

$$X \leqslant \Psi$$

is by our definition of $\leqslant$ equivalent to

$$(\forall r)((\forall s)(\Psi s\rightarrow s \leqslant r)\rightarrow(\forall s)(Xs\rightarrow s \leqslant r))$$

and when X is defined as above this is to say that

$(\forall r)((\forall s)(\Psi s\rightarrow s \leqslant r)$
$$\rightarrow(\forall s)((\exists\Phi)(Bounded(\Phi)\ \&\ (\mho r)(\Phi r)\ \&\ \Phi s)\ \rightarrow s \leqslant r)).$$

By elementary quantifier rules this is equivalent to

$(\forall\Phi)(Bounded(\Phi)\ \&\ (\mho r)(\Phi r)$
$$\rightarrow(\forall r)((\forall s)(\Psi s\rightarrow s \leqslant r)\rightarrow(\forall s)(\Phi s\rightarrow s \leqslant r)))$$

i.e. to

$$(\forall\Phi)(Bounded(\Phi)\ \&\ (\mho r)(\Phi r)\ \rightarrow \Phi \leqslant \Psi).$$

This completes the proof.

It may be observed that throughout the proof the schema $(\Omega r)(\Phi r)$ has been *idle*, in that we have made no use of the rules governing the symbol Ω, but we do of course presume that it represents any arbitrary predicate of predicates of rational numbers when we say that the theorem just established *is* the principle of least upper bounds for these predicates (and we shall later be making use of this presumption (pp. 255–6)). We have presupposed a thesis of second-order quantification, reflecting the fact that predicates of rational numbers are indeed predicates (remainders), in assuming 'impredicatively' that

$$(\exists X)(\forall s)(Xs \leftrightarrow (\exists \Phi)(Bounded(\Phi) \ \& \ (\Omega r)(\Phi r) \ \& \ \Phi s)).$$

Taking into account our understanding of Φ and X in this use, this is the assumption that

$$(\exists \Re)(\forall pq)(\exists p \Re \exists q \leftrightarrow (\exists \mathfrak{S})(Bounded(\mathfrak{S})$$
$$\& \ (\Omega nm)(\exists n \mathfrak{S} \exists m) \ \& \ \exists p \mathfrak{S} \exists q))$$

and it is a consequence of the substitution rules adopted originally for predicates of pure quantifiers when these are extended in an obvious way to cover dyadic predicates. I would, then, think it perfectly reasonable to claim that we can show as a matter of 'pure logic' that there is an unbounded and continuous quantity[35].

Notice that for the purposes of this proof it is quite irrelevant whether our account of 'predicates of rational numbers' does adequately capture our intuitive understanding of this notion, nor indeed does it matter whether the numerical quantifiers that form the basis of the argument are or are not to be identified with the natural numbers. For all that is being claimed is that there is a continuous and unbounded quantity, and how that quantity is related to numbers is of no importance. However it is important that this quantity should come within the range of our quantification over all extensive quantities, i.e. within the range of our expression 'for all subscripted ϕ' as elucidated on pp. 190–1. Thus the type-neutrality of these postulates, and of our quantification over all relations that satisfy them, is crucial. This is because the further development of the theory of ratios will require us to prove that there

---

[35] See Volume 1, Chapter 6, §1.

does exist a continuous quantity in the sense defined in that theory. Let us turn, then, to this further development.

Instead of taking the route used in §2, whereby we would now turn directly to showing that the ratios themselves may be viewed as instances of a continuous quantity, I shall this time develop the theory in close parallel to the development of the theory of real numbers in §1. To bring out this parallel, let us now use the same variables N, M, P, . . . to stand in directly for ratios, so that

$$(\forall N)(-N-)$$

will be understood simply as an alternative way of writing[36]

$$(\forall\phi)(\forall xy)(-(x:_\phi y)-).$$

Now the relation $\leqslant$ between ratios is easily shown to be a total ordering, so that we have

$$(\forall NM)(N\leqslant M \vee M\leqslant N)$$

$$(\forall NM)(N\leqslant M \ \& \ M\leqslant N \to N=M)$$

$$(\forall NMP)(N\leqslant M \ \& \ M\leqslant P \to N\leqslant P).$$

Furthermore, it is a dense ordering without first or last members and contains a denumerable and relatively dense subset. For from the definition of the variable N we evidently have

$$(\exists N)(N=n:m)$$

the rational ratios $n:m$ are of course denumerable, and we can prove without difficulty

$$(\exists nm)(n:m < N)$$

$$(\exists nm)(n:m > N)$$

$$N < M \to (\exists nm)(N < n:m \ \& \ n:m < M).$$

The character of the ordering will therefore be completely

---

[36] To be more exact, we should add the condition $Sat(x, y, \phi)$, as elucidated on p. 191, and rewrite with type-neutral variables to obtain the definition

$$(\forall N)(\mathscr{A}(N)) \quad \text{for} \quad (\forall\phi)(\forall\alpha\beta)(Sat(\alpha, \beta, \phi)\to\mathscr{A}(\alpha:_\phi\beta))$$

provided that (i) every occurrence of α, β, and ϕ in the formula $\mathscr{A}(\alpha:_\phi\beta)$ is in the context $\alpha:_\phi\beta$, and (ii) N occurs in $\mathscr{A}(N)$ wherever and only where $\alpha:_\phi\beta$ occurs in $\mathscr{A}(\alpha:_\phi\beta)$. Similarly for alphabetic variants.

established[37] if we can show that the principle of least upper bounds holds for ratios.

The simplest plan here is to proceed again by correlating the ratios with instances of a quantity already known to be continuous and unbounded, namely (in our case) with the *bounded*[38] predicates of rational numbers. We already know, from the theorem of the existence of the fourth proportional, that for any bounded predicate X we have

(i) $(\forall N)(\exists \Phi)(Bounded(\Phi) \ \& \ N = \Phi : X))$

and since ratios between bounded predicates of rational numbers are indeed ratios we have conversely

(ii) $(\forall \Phi)(Bounded(\Phi) \rightarrow (\exists N)(N = \Phi : X)).$

Therefore, given any bounded predicate X as base for the correlation, we know that every ratio can be thus correlated with a bounded predicate of rational numbers, and every bounded predicate of rational numbers can be thus correlated with a ratio. Furthermore, it is obvious that this correlation preserves the ordering in that

(iii) $N = \Phi : X \ \& \ M = \Psi : X \rightarrow (N \leqslant M \leftrightarrow \Phi \leqslant \Psi).$

It therefore follows that L.U.B. for bounded predicates of rational numbers will transfer *via* the correlation to L.U.B. for ratios themselves.

The detail of the proof is perfectly straightforward, though in order to *state* the conclusion we are aiming for we must once more introduce a new variable, say $\mathscr{F}$, for predicates of ratios. I shall say something of this shortly, but for the moment let us assume it done. Then we begin by assuming our antecedent

$$(\exists N)(\mathscr{F}N) \ \& \ (\exists M)(\forall N)(\mathscr{F}N \rightarrow N \leqslant M).$$

Confining the quantifiers to *bounded* predicates of rational numbers for brevity, it is then obvious from (i) and (ii) that for any bounded predicate X we have

$$(\exists \Phi)(\exists N)(\mathscr{F}N \ \& \ N = \Phi : X)$$

$$(\exists \Psi)(\forall \Phi)((\exists N)(\mathscr{F}N \ \& \ N = \Phi : X) \rightarrow \Phi \leqslant \Psi).$$

---

[37] See footnote 13, p. 208.

[38] For brevity I henceforth use 'bounded' in the text, like '*Bounded*' in the formulae, as short for 'non-empty and bounded above'.

Now to these premises we apply L.U.B. for bounded predicates of rational numbers, with the formula $(\exists N)(\mathscr{F}N \ \& \ N = \Phi : X)$ in place of the elementary schema $(\Omega r)(\Phi r)$, and this is perfectly legitimate because when this formula is expanded into more primitive notation the occurrences of Φ are of course all in the context Φr. Thus we infer that there is some bounded predicate Ω of rational numbers such that

$$(\forall \Psi)(\Omega \leqslant \Psi \leftrightarrow (\forall \Phi)((\exists N)(\mathscr{F}N \ \& \ N = \Phi : X) \rightarrow \Phi \leqslant \Psi)).$$

Hence there follows, by (iii) above,

$$(\forall \Psi)(\Omega : X \leqslant \Psi : X \leftrightarrow (\forall \Phi)((\exists N)(\mathscr{F}N \ \& \ N = \Phi : X)$$
$$\rightarrow \Phi : X \leqslant \Psi : X))$$

i.e.

$$(\forall \Psi)(\Omega : X \leqslant \Psi : X \leftrightarrow (\forall N)((\exists \Phi)(N = \Phi : X) \ \& \ \mathscr{F}N$$
$$\rightarrow N \leqslant \Psi : X)).$$

In view of (i) again this may be simplified to

$$(\forall \Psi)(\Omega : X \leqslant \Psi : X \leftrightarrow (\forall N)(\mathscr{F}N \rightarrow N \leqslant \Psi : X))$$

and then further to

$$(\forall M)(\Omega : X \leqslant M \leftrightarrow (\forall N)(\mathscr{F}N \rightarrow N \leqslant M))$$

and so finally in view of (ii) to

$$(\exists P)(\forall M)(P \leqslant M \leftrightarrow (\forall N)(\mathscr{F}N \rightarrow N \leqslant M))$$

which is the desired result. Any *instance* of this theorem, with a specific predicate of ratios in place of $\mathscr{F}$, is of course obtainable by this method without a prior introduction of the variable $\mathscr{F}$, because again no particular properties of $\mathscr{F}$ were assumed in the proof. So the introduction of this new variable plays no important part in the formal development of the theory of ratios but it is needed for the general statement of our theorem, and this time the proper introduction of the variable is not quite so straightforward. I shall shortly return to it.

From here the next step will be to introduce addition and multiplication of ratios. In §2 we first suggested that addition of ratios could be introduced by correlating the ratios with the instances of some quantity already assumed to be continuous, for example length, because this method makes it obvious that the addition so introduced has the required properties. The particular choice of *length* would evidently be unfortunate in a *general* development of the theory of ratios, because we would not want the whole theory to depend on specifically geometrical axioms about length. This drawback is of course avoided if we change our example to any quantity that can be proved to be continuous as a matter of logic, and so we might once more use our bounded predicates of rational numbers. We would then single out some one such predicate X to be the basis of the correlation, and define

$$N + M = P$$

for

$$(\exists \Phi \Psi)(N = \Phi : X \ \& \ M = \Psi : X \ \& \ P = \Phi \oplus \Psi : X).$$

Since it does not matter which predicate is chosen as the basis, we could generalize this definiens by adding an initial quantifier $(\exists X)$, or we could generalize still further to

$$(\exists \phi)(\exists xyz)(N = x :_\phi z \ \& \ M = y :_\phi z \ \& \ P = x \oplus y :_\phi z).$$

As I showed earlier, these variations do not affect the definition, and as a matter of fact the main properties of addition would be forthcoming from the last definition even if there were no continuous quantity. However, there is no necessity to proceed in this way, and if it was thought preferable we could instead copy the method used for addition of real numbers in §1. Here we would rely on our previous knowledge of how to add *rational* ratios (p. 188), and understand $N + M$ as the ratio which is the least upper bound of all the rational ratios $(n:m) + (p:q)$ such that $n:m \leqslant N \ \& \ p:q \leqslant M$. On any of these methods we shall prove without difficulty that addition has the expected properties, and hence (since L.U.B. is already available) that the ratios themselves may be viewed as instances of a quantity that is continuous and unbounded.

The definition of multiplication offers the same alternatives as that of addition. We may follow the first suggestion of §2, introduce ratios of ratios, and regard

$$N \cdot M = P$$

as a fairly trivial rewriting of

$$N = P : M.$$

Alternatively we may proceed in closer analogy to the definition of addition in §2, and give as definiens

$$(\exists\phi)(\exists xyz)(N = x:_\phi y \,\&\, M = y:_\phi z \,\&\, P = x:_\phi z).$$

Or again we may use the method of §1 and understand $N \cdot M$ as the least upper bound of all the rational ratios $(n:m) \cdot (p:q)$ such that $n:m \leqslant N \,\&\, p:q \leqslant M$.

These definitions allow us to set up an *algebra* of ratios precisely parallel to the usual algebra for positive real numbers. So far we have provided the means for establishing

$$N + M = M + N$$
$$N + (M + P) = (N + M) + P$$
$$N < M \leftrightarrow (\exists P)(N + P = M)$$
$$N \cdot M = M \cdot N$$
$$N \cdot (M \cdot P) = (N \cdot M) \cdot P$$
$$N \cdot (M + P) = (N \cdot M) + (N \cdot P).$$

We could go on to introduce 1 now as the name of a ratio, namely the ratio of anything to itself, and prove that this was a unit ratio with respect to multiplication:

$$N \cdot 1 = N.$$

We could also go on to show that the operation of multiplication had a unique inverse, perhaps by introducing the reciprocal N^{-1} of any ratio N (so defined that if $N = a:b$ then $N^{-1} = b:a$), and proving that

$$N \cdot N^{-1} = 1.$$

It is obvious that there is a complete isomorphism between our system of ratios and the system of real numbers as classically conceived, and any characteristics of the one will reappear in the other. So now let us come to the question whether we can in fact simply *identify* the real numbers with the ratios.

If real numbers and ratios may simply be identified, the theory of proportion we have been developing becomes a construction of the theory of real numbers to rival the 'classical' constructions discussed in §1. Now Dedekind and his contemporaries may well have seen that the use which geometry makes of proportions contains all that is essential to the concept of a real number, but I suspect that they shied away from a direct use of the geometrical notion of proportion mainly because they failed to disentangle the *general* theory of proportion from its specifically geometrical applications, and consequently it seemed to them that this notion depended on specifically geometrical axioms. If this were true, then it would be quite understandable that they avoided the Greek theory, for after the advent of non-Euclidean geometry it had become clear that geometrical axioms had no strong claim to be *a priori* truths, whereas the theory of real numbers did seem to them to be *a priori*.

A nice illustration of this point can be seen in Dedekind's complaint[39] that until his time no one had produced any satisfactory proof of the thesis

$$\sqrt{2} \cdot \sqrt{3} = \sqrt{6}$$

for it is very easy to give a *geometrical* proof of this thesis from Euclidean axioms. For example we may rephrase the thesis as

$$\sqrt{1} : \sqrt{2} = \sqrt{3} : \sqrt{6}$$

and we may then interpret the ratio $\sqrt{n} : \sqrt{m}$ geometrically as the ratio in length between the sides of squares which have a ratio in area of $n : m$. Given this interpretation, *and* of course the Euclidean axioms, a proof is forthcoming at once. However, one can well understand that this would seem to Dedekind and his contemporaries a questionable basis for what is not in itself a specially geometrical theorem.

The chief motive, then, for the endeavour to 'arithmetize' analysis was the thought that it should certainly be 'de-geometrized', i.e. it was quite improper to let the theory of real numbers rest on geometrical axioms or geometrical intuition. This thought we can certainly agree with, if only because the real numbers may properly be used in connection with *any* continuous quantity, and not only the geometrical ones. However, there is no strong reason to agree with the positive contention that the theory of real numbers should be constructed directly from that of the natural numbers. This manner of proceeding ignores all application of real numbers to quantities

[39] Dedekind 1872, p. 22.

and in consequence it can easily lead to quite inadequate views on such applications. For example if the real numbers are pictured as existing in splendid mathematical isolation it becomes a problem to see how their application to quantities can be anything more than an entirely arbitrary use of the sort of 'measuring labels' discussed in Chapter 2, § 2, and moreover we lose any reason beyond mere linguistic economy for using these same real numbers in application to many different quantities. To put the point briefly, real numbers are in the first place demanded by the existence of continuous quantities, and in particular by the fact that ratios in respect of *different* quantities are yet comparable with one another. However, since the arithmetizing approach has nothing to say about quantities it does not of itself provide any satisfactory explanation of why we should be interested in real numbers, or of what use they have. Of course this is not to say that it actually *prevents* us *adding* such explanations, but when we do we find that (*if* real numbers and ratios can be identified) the explanation itself furnishes a construction of the real number system.

However, it would not be fair to conclude that the arithmetizing approach is thus shown to be entirely superfluous, for it has its role to play *within* the approach based on the theory of proportion. We have seen that in order to develop the theory to a point where it can be claimed to be an adequate construction of the real numbers it was necessary to show that there did exist a continuous quantity. (Similarly in constructing the theory of natural numbers by purely extensional methods it is necessary to show that there does exist a denumerable totality, whether of the numbers themselves or of something else.) Now the quantities to which real numbers are ordinarily applied,— for example mass, length, time—are standardly assumed to be continuous, but the assumption would appear to be an empirical one and at any rate not the sort of assumption that ought to lie at the basis of the theory of real numbers. We found it necessary, therefore, to construct a somewhat artificial quantity which *could* be proved to be continuous, and here we frankly took over a construction from the arithmetizing approach. However, the difference was that this construction was used simply to provide an *example* of a continuous quantity, and the fact that the several alternative arithmetizing approaches furnish several alternative ways of proving this required existential theorem is evidently not an objection to the present approach. Our theory remains a general theory of *all* extensive quantities, though naturally it does not ignore the fact that the ratios may themselves be viewed as instances of an extensive quantity. As for the reliance on geometrical or other special axioms, it should be noted that our theory does not formally *depend* on the postulates for extensive quantities in the sense that these are taken as premises to the general theorems about ratios that yield real number theory. For these theorems are all conditional in form and state (roughly) that *if* any quantity satisfies the postulates then certain further results hold for it. Of course what makes the theory interesting is that, as a matter of empirical fact, many physically significant quantities do appear to satisfy the postulates, but the theory is not itself construed as an empirical theory.

The considerations of these last paragraphs are intended to show that, if we wish to construct (rather than postulate) the real numbers, there is good reason to make the construction go *via* the concepts of the general theory of proportion. However, they do not also require us to finish the construction by simply *identifying* the real numbers with the ratios, and in fact that identification is open to the objection we raised earlier to the proposal that the rational numbers be identified with the rational ratios, and to one of the objections we raised in §1 to all classical constructions. On this proposal no natural number can *be* a real number, but our ordinary view surely is that the natural numbers are themselves amongst the real numbers; the natural number 3 is ordinarily not distinguished from the real number 3 but it certainly is distinguished from the ratio 3:1, and this is shown by the fact that we do not scruple to add a natural number to a real number to form a further number but we surely are not very happy to add together a natural number and a ratio.

When this objection was raised earlier against the proposal to identify rational numbers with ratios, we sought to avoid it by deliberately introducing the usual symbols for rational numbers in a context in which numerals for natural numbers also appear, *viz.* in statements of numerical comparison, and then regarding the theory of rational numbers as abstracted from *this* use of rational numbers. If we now wish to pursue the same plan for real numbers, it might seem that all we have to do is to allow ourselves to rewrite $a:b = N$ as $a \approx_N b$, which we could then compare appropriately with $a \approx_{(n/m)} b$ and $a \approx_n b$. However, in the present case this simple manoeuvre is surely altogether too simple to achieve the desired result, for it must presumably follow that $a \approx_{(x:\varphi y)} b$ is a well-formed formula, whereas we had agreed that ratios were not themselves numbers and were wishing to preserve the position of N in $a \approx_N b$ as a position only available to expressions for numbers[40].

Let us recant, therefore, the over-hasty use of the variables N, M, ..., to stand in place of expressions for ratios, and consider whether we can introduce these variables for the first time in contexts of the form $a \approx_N b$, where we shall now have to add an explicit subscript ϕ. Our problem, then, is to say when a proposition should be said to be of the form $a \approx_{N.\varphi} b$, and I confess that I do not think that this problem can be adequately resolved with our present apparatus. One difficulty is that to obtain full generality we ought really to encompass numerical comparisons in respect of quantities which are not extensive quantities, for example density or velocity, but perhaps these could be handled as secondary in the manner suggested on pp. 243–4. For simplicity, then, let us confine our attention to extensive quantities, where we can view the subscript ϕ in $a \approx_{N.\varphi} b$ as a variable for those homogeneous triadic relations that are capable of satisfying our postulates for extensive quantities (p. 156.

One thing that is clear is that propositions of the form $a \approx_{N.\varphi} b$ do give information about the ratio $a:_\varphi b$, and so we may first consider whether we can give a general characterization of propositions about ratios, i.e. of propositions of the form $\mathscr{F}(a:_\varphi b)$, thereby introducing the variable $\mathscr{F}$ used earlier. We may begin from the observation that any proposition about the ratio $a:_\varphi b$ can always be viewed as asserting that a certain relationship holds between a and b and ϕ-ness, so any such statement has the general form $\mathfrak{R}(a, b, \phi)$. Furthermore, if $\mathfrak{R}(a, b, \phi)$ is indeed a statement about the ratio $a:_\varphi b$, then it

---

[40] By contrast, in Chapter 2 the contexts in which $n:m$ occurred were available *only* to ratios between natural numbers, so when we introduced $\approx_{(n/m)}$ we did not licence the introduction of $\approx_{(x:y)\varphi}$.

must be possible to make that same statement about any other ratio, so $\Re$ must be sufficiently type-neutral to apply significantly to any other trio (c, d, ψ) where ψ is a homogeneous triadic relation of any type that satisfies our postulates and c and d are in its field. We must therefore invoke a type-neutrality more complex than any explicitly discussed so far, though the general principles on which we earlier proceeded (in Volume 1, Chapter 4) can doubtless be extended to cover this sort of case. The simplest plan is then to stipulate that $\Re$ should be such that $\Re(a, b, \phi)$ is significant wherever ϕ is a homogeneous triadic relation of any type, and a and b are of such a type as to be significant terms to the relation. We could require in addition that ϕ does actually satisfy our postulates, and that a and b are actually within its field (i.e. we could require that $Sat(a, b, \phi)$ in the sense of p. 191), but I doubt whether this would be desirable. For example it might turn out that the quantity mass did *not* satisfy our postulates, and then with this ruling a proposition stated in the form 'the ratio of a to b in mass is 2 to 1' would *not* be a proposition of the form $\mathscr{F}(a :_{\text{mass}} b)$, which seems to be the wrong result.

A weaker and more plausible criterion is that if a proposition is to be of the form $\mathscr{F}(a :_{\varphi} b)$ then it must at least *imply* $Sat(a, b, \phi)$, but even this is not entirely straightforward. A crucial case would be what we wish to say about 'there is no such thing as the ratio of a to b in mass' (asserted perhaps on the ground that $\sim Sat(a, b, \phi)$) for this proposition does appear to be about a ratio but it evidently does not imply that there is such a ratio. Here the natural response is to remark that expressions such as 'the ratio of a to b in mass' are from a linguistic point of view definite descriptions, and therefore they should be assigned *scopes* like all other definite descriptions even though we are not regarding them as uniquely describing any *objects*. This means in effect that we do not give any one set of conditions which all propositions about $a :_{\varphi} b$ have to satisfy, but we proceed in two stages. First we single out the propositions in which the description has major scope; these all do imply that the ratio exists, and may perhaps be called *primary* propositions of the form $\mathscr{F}(a :_{\varphi} b)$. Then we add that there are other secondary propositions which may also be said to be of this form but which lack the existential implication, and these are more complex propositions which are built up by truth-functional combination or other methods from primary propositions of the form $\mathscr{F}(a:_{\varphi}b)$. However, for present purposes it will do no harm if we concentrate exclusively on the primary cases and therefore do impose the existential condition.

Another characteristic of primary propositions of the form $\mathscr{F}(a:_{\varphi}b)$ is that they certainly will not include any intentional contexts for $a:_{\varphi}b$, which is to say that if $a:_{\varphi}b$ and $c:_{\psi}d$ are the same ratio then $\mathscr{F}(a:_{\varphi}b)$ and $\mathscr{F}(c:_{\psi}d)$ must have the same truth-value. I do not think any other conditions are required, and therefore it seems to me adequate to stipulate that a proposition is a primary proposition of the form $\mathscr{F}(a:_{\varphi}b)$ if and only if it is of the general form $\Re(a, b, \phi)$ where (i) $\Re$ is appropriately type-neutral and (ii) $\Re$ satisfies the two conditions

$$(\forall\phi)(\forall xy)(\Re(x, y, \phi) \rightarrow Sat(x, y, \phi))$$

$$(\forall\phi\psi)(\forall xyzw)(Sat(x, y, \phi) \ \& \ Sat(z, w, \psi) \ \& \ x:_{\varphi}y = z:_{\psi}w$$
$$\rightarrow (\Re(x, y, \phi) \leftrightarrow \Re(z, w, \psi))).$$

If we abbreviate the conjunction of these two conditions to $SAT(\Re)$, then from a formal point of view the variable $\mathscr{F}$ may be introduced in such a way that

$$(\forall\mathscr{F})(\text{---}\mathscr{F}(a:_{\varphi}b)\text{---})$$

is understood simply as abbreviating

$$(\forall\Re)(SAT(\Re) \rightarrow (\text{---}\Re(a, b, \phi)\text{---}))$$

provided that $\Re$ has already been introduced as a variable for type-neutral relations of the appropriate sort. I shall not delay any further with the type-neutrality requirement, because as I say I think that our ultimate goal will elude us.

The ultimate goal was to determine when a proposition should be said to be of the form $a \approx_{N.\varphi} b$. Since these propositions evidently do give information about the ratio $a:_{\varphi}b$ and imply that this ratio exists, we may certainly start with the requirement that the proposition be of the form $\mathscr{F}(a:_{\varphi}b)$. Another requirement emerges from the observation that if $a \approx_{N.\varphi} b$ is true, then $c \approx_{N.\psi} d$ is true also if and only if

$a:_\varphi b = c:_\psi d$, so we may lay it down that $\mathscr{F}$ should generate an equivalence class among proportions, i.e. that

$$(\exists\phi)(\exists xy)(Sat(x, y, \phi) \,\&\, (\forall\psi)(\forall zw)(Sat(z, w, \psi)$$
$$\rightarrow(\mathscr{F}(z:_\psi w) \leftrightarrow z:_\psi w = x:_\varphi y))).$$

If these conditions form an adequate analysis we may then introduce the formula $a \approx_{N.\varphi} b$ by regarding

$$(\forall N)(-\!\!-(a \approx_{N.\varphi} b)-\!\!-)$$

simply as an abbreviation for

$$(\forall\mathscr{F})((\exists\psi)(\exists xy)(Sat(x, y, \psi) \,\&\, (\forall\chi)(\forall zw)(Sat(z, w, \chi)$$
$$\rightarrow(\mathscr{F}(z:_\chi w) \leftrightarrow z:_\chi w = x:_\psi y))) \rightarrow (-\!\!-\mathscr{F}(a:_\varphi b)-\!\!-)).$$

Using the fact, secured by the introduction of $\mathscr{F}$, that

$$(\forall\mathscr{F})(\forall\chi)(\forall zw)(\mathscr{F}(z:_\chi w) \rightarrow Sat(z, w, \chi)).$$

We can rewrite this as

$$(\forall\mathscr{F})(\forall\psi)(\forall xy)(Sat(x, y, \psi) \,\&\, (\forall\chi)(\forall zw)(\mathscr{F}(z:_\chi w)$$
$$\leftrightarrow (z:_\chi w = x:_\psi y \,\&\, Sat(z, w, \chi))) \rightarrow (-\!\!-\mathscr{F}(a:_\varphi b)-\!\!-))$$

and hence as

$$(\forall\psi)(\forall xy)(\forall\mathscr{F})(Sat(x, y, \psi) \,\&\, (\forall\chi)(\forall zw)(\mathscr{F}(z:_\chi w)$$
$$\leftrightarrow (z:_\chi w = x:_\psi y \,\&\, Sat(z, w, \chi)))$$
$$\rightarrow (-\!\!-(a:_\varphi b = x:_\psi y \,\&\, Sat(a, b, \phi))-\!\!-))$$

and so once more as

$$(\forall\psi)(\forall xy)(Sat(x, y, \psi) \,\&\, (\exists\mathscr{F})(\forall\chi)(\forall zw)(\mathscr{F}(z:_\chi w)$$
$$\leftrightarrow (z:_\chi w = x:_\psi y \,\&\, Sat(z, w, \chi)))$$
$$\rightarrow (-\!\!-(a:_\varphi b = x:_\psi y \,\&\, Sat(a, b, \phi))-\!\!-).$$

However, the second antecedent is now superfluous, since it may be trivially satisfied by taking $\mathscr{F}(z:_\chi w)$ precisely as $z:_\chi w = x:_\psi y \,\&\, Sat(z, w, \chi)$ and so we come down to

$$(\forall\psi)(\forall xy)(Sat(x, y, \psi) \rightarrow (-\!\!-(a:_\varphi b = x:_\psi y \,\&\, Sat(a, b, \phi))-\!\!-))$$

as our equivalent for

$$(\forall N)(-\!\!-(a \approx_{N.\varphi} b)-\!\!-).$$

Our new variables N are therefore introduced in such a way that a formula containing them is again equivalent to a formula of the general theory of proportion, but now the equivalence is supposed to be justified by an analysis of what it is for a proposition to be of the form $a \approx_{N\varphi} b$.

Unfortunately the analysis consisting just of the conditions we have mentioned is inadequate, because it does not incorporate the point that the information which $a \approx_{N.\varphi} b$ gives about the ratio $a :_{\varphi} b$ specifically concerns the *numerical* value of that ratio. The analysis we have offered evidently permits us to count $a :_{\varphi} b = c :_{\psi} d$ as itself a proposition of the form $a \approx_{N.\varphi} b$, though this proposition need not give us any information about the numerical value of $a :_{\varphi} b$ unless the numerical value of $c :_{\psi} d$ is already known or at least determinable. Without further information we cannot, simply from $a :_{\varphi} b = c :_{\psi} d$, determine the truth or falsity of any proposition of the form $a :_{\varphi} b \geqslant n : m$, though it seems clear that every such proposition is automatically determined by any proposition of the form $a \approx_{N.\varphi} b$[41]. If we permit ourselves to introduce a necessity-operator we might try to meet this objection by adding an extra condition

$$(\forall nm)(\Box(\forall\phi)(\forall xy)(\ \mathscr{F}(x :_{\varphi} y) \rightarrow x :_{\varphi} y \geqslant n : m)$$
$$\vee\ \Box(\forall\phi)(\forall xy)(\mathscr{F}(x :_{\varphi} y) \rightarrow x :_{\varphi} y < n : m))$$

but we should then need to give a proper discussion of how the necessity-operator is to be understood and of what rules are to govern its employment, which I do not at present feel able to give. A further, and quite noticeable, difficulty would be the question of how to prove the existence of a predicate $\mathscr{F}$ satisfying this condition for each real number[42]. Besides, even if

[41] I am now inclined to think that the introduction of numerical quantifiers which I gave in Volume 1 (Chapter 5, §2) is open to the same objection, and that it is not after all fully protected by the stipulation that the numerical quantifiers are to be type-neutral. *Perhaps* the addition of a necessity-operator there *would* 'improve matters' (Volume 1, pp. 157–8), but the point needs more discussion than I can give it here. See also pp. 279–82.

[42] Recall that I distinguish predicates from predicate-expressions, so the denumerability of the latter is no objection. Predicates exist just in case propositions containing them exist, and propositions exist wherever there are truths, for they are required to be what it is that is true. However, there are certainly non-denumerably many truths, because for example there is a truth concerning each real number to the effect that it is a real number. See Volume 1, pp. 75–6.

these obstacles could be overcome there is still the rather fundamental objection that we do not seem to have got very far
towards providing a convincing answer to the question: What,
on this account, is supposed to be the difference between the
real number N and the ratio $N:1$?

In default of any adequate analysis of propositions employing real numbers, the best we can do is to argue first that for
any real number N there is always a ratio $x:_{\psi}y$ (if only between
bounded predicates of rational numbers) such that[43]

$$a \approx_{N \cdot \varphi} b \leftrightarrow a:_{\varphi}b = x:_{\psi}y.$$

Conversely it is also clear that for any ratio $x:_{\psi}y$ there is
always a real number N such that

$$a \approx_{N \cdot \varphi} b \leftrightarrow a:_{\varphi}b = x:_{\psi}y.$$

If this is granted, it then follows that the formulae

$$(\forall N)(-(a \approx_{N \cdot \varphi} b)-)$$

and

$$(\forall \psi)(\forall xy)(-(a:_{\varphi}b = \ x:_{\psi}y)-)$$

will always be equivalent (provided, as will be the case for all
the examples we are actually concerned with, that the relevant
context is extensional). Therefore we commit no error in taking
the one formula to be interchangeable with the other, but if
we use this point to introduce variables for real numbers by
definition, the claim that the definition is adequate is a claim
for which no very satisfactory argument has been given. If the
claim is granted, there is no difficulty in giving a construction
of the theory of real numbers from this starting point. For
example we could more generally define

$$(\forall N)(-(a \succcurlyeq_{N \cdot \varphi} b)-)$$

as an abbreviation for

$$(\forall \psi)(\forall xy)(-(a:_{\varphi}b \geqslant x:_{\psi}y)-).$$

Then we could go on to define

$$N \geqslant M \quad \text{for} \quad (\forall \phi)(\forall xy)(x \succ_{N \cdot \varphi} y \to x \succ_{M \cdot \varphi} y)$$

[43] For brevity I henceforth omit the qualifying clauses $Sat(a, b, \phi)$ etc.
(cf. footnote 36, p. 254).

just as we did in the case of rational numbers. Further definitions offer no more difficulty here than they did there.

The only advantage of this proposal over the previous one is that by making use of such formulae as $a \succcurlyeq_N b$, and thereby suggesting comparison with $a \succcurlyeq_{(n/m)} b$ and $a \succcurlyeq_n b$, it makes it more natural to see the natural numbers, the rational numbers, and the real numbers as all fitting in to one comprehensive system. However, since nothing very much is said about the *meaning* of $a \succcurlyeq_N b$ this may not seem a very substantial gain, and if we take it that the meaning is to be given by the definition there is no appreciable gain at all. In each case claims about real numbers will be taken to mean the same as claims about ratios, and though the claims will be slightly different in the two cases there is nothing very much to choose between them. For example if we consider the thesis   *So by the same argument ...*

$$N \geqslant M \vee M \geqslant N$$

on our second proposal this is to be identified with the thesis

$$(\forall \phi)(\forall xy)(x \succcurlyeq_{N \cdot \varphi} y \to x \succcurlyeq_{M \cdot \varphi} y) \vee (\forall \phi)(\forall xy)(x \succcurlyeq_{M \cdot \varphi} y \to x \succcurlyeq_{N \cdot \varphi} y)$$

and this in turn is to be identified with

$$(\forall \phi)(\forall xy)(x :_\varphi y \geqslant a :_\psi b \to x :_\varphi y \geqslant c :_\chi d)$$
$$\vee (\forall \phi)(\forall xy)(x :_\varphi y \geqslant c :_\chi d \to x :_\varphi y \geqslant a :_\psi b).$$

Now it is easily seen that

$$a :_\psi b \geqslant c :_\chi d \leftrightarrow (\forall \phi)(\forall xy)(x :_\varphi y \geqslant a :_\psi b \to x :_\varphi y \geqslant c :_\chi d).$$

So on our second proposal the original thesis about real numbers is to be. identified with a thesis easily seen to be *equivalent* to

$$a :_\psi b \geqslant c :_\chi d \vee c :_\chi d \geqslant a :_\psi b$$

and this latter thesis is what we would have *identified* with

$$N \geqslant M \vee M \geqslant N$$

according to our first proposal by which real numbers were taken to *be* ratios. On either proposal statements apparently about real numbers will translate into statements about ratios, and the two translating formulae will easily be shown to be interdeducible.

At this point we may perhaps go on to add that if we trace

the construction back further the statements apparently about ratios *need* not be seen as concerned with any such things as ratios, in so far as

$$a :_\psi b \geqslant c :_\chi d$$

is translated into

$$(\forall nm)(a :_\psi b \geqslant n : m \rightarrow c :_\chi d \geqslant n : m)$$

and this in turn into

$$(\forall nm)\left((\exists pq)\left(\frac{1}{p} \circ a \succcurlyeq_\psi \frac{1}{q} \circ b \ \& \ p : q \geqslant n : m \right) \right.$$
$$\left. \rightarrow (\exists pq)\left(\frac{1}{p} \circ c \succcurlyeq_\chi \frac{1}{q} \circ d \ \& \ p : q \geqslant n : m \right) \right)$$

and this again may be further translated by applying our other definitions, so that every one of our statements is in the end concerned solely with comparative relations and natural numbers (occurring, say, as powers of relations). From the purely formal point of view the whole construction is *syncategorematic*, in the sense that no new entities are introduced beyond those involved in the original postulates, and statements apparently concerning real numbers, or ratios, or rational numbers, or fractions are all introduced as abbreviations for statements which have no such concern. But then, as I argued earlier with respect to our first construction of natural numbers[44], if in this way we treat the whole construction as *just* a series of purely logical deductions it is of no philosophical relevance whatever. What makes the construction philosophically interesting are such extra claims as that the symbol $a :_\varphi b$ as it occurs in the construction does adequately represent our notion of a ratio. Such claims may be supported either on the ground that the formal definition of the symbols assigns them precisely the meaning which they are ordinarily understood to have, or on the ground that when we analyse the meaning they are ordinarily understood to have we can thereby show that what is given as a definition must at any rate be equivalent in truth-value. I have to admit that I cannot really support such a claim for our use of the variable N even as it occurs in the formula $a \succcurlyeq_{N.\varphi} b$. No doubt it is perfectly correct to claim that our way

44 Volume 1, Chapter 6, §§ 2, 3.

of defining this formula must preserve truth-values, but to *justify* this claim one must be able to *give* the analysis of the formula, which I have not yet succeeded in doing.

When we turn to consider the move from $a \succcurlyeq_{N.\varphi} b$ to $N \geqslant M$ there is of course yet more reason for doubt. The general strategy of our approach has been to consider the way in which real numbers are applied to extensive quantities and then to construe the arithmetic of real numbers as obtained by a sort of abstraction from those applications, the idea being that apparently singular statements about real numbers are taken as really being general statements about all applications of those numbers. If this strategy is to succeed, we must somehow assure ourselves that the applications considered are either the only or at any rate the most fundamental applications. However, although the particular type of application we have singled out was surely the most important from a *historical* point of view, it would certainly not be easy to argue that the same is true today, and I shall not make any attempt to do so.

Appendix. On Postulates for the Theory of Proportion

The Greek theory of proportion due to Eudoxus, and preserved for us in Euclid, Book V, differs from the theory presented in this book in two ways. My postulates (1')–(6') do not include the assumption that every quantity is unbounded, whereas Euclid and virtually all his successors do include it[49]; and my postulates do include the principle of least upper bounds, which is lacking in Euclid and in some contemporary writers. This principle is needed for Euclidean geometry (as traditionally conceived) and will certainly be assumed nowadays for geometrical quantities. (Euclid's lack of the principle renders several of his proofs invalid, e.g. Book XII, Propositions 2, 5, 11, 12, 18.) However, some contemporary writers prefer not to assume it for *all* extensive quantities.

Roughly speaking, the assumption that every quantity is unbounded is replaced in my treatment by the theorem (p. 173) that every quantity is either discrete or infinitely divisible. The assumption of unboundedness is certainly more convenient. It allows us to simplify definitions (as on p. 185–6) and permits simple methods of proof where my version has to proceed more circuitously. To take a simple example, consider the elementary theorem

$$n \circ a \succ p \circ b \ \& \ m \circ a \approx q \circ b \rightarrow n \cdot q > m \cdot p.$$

---

[45] An exception is the unusual treatment sketched by Suppes (1951, pp. 4–8).

Euclid would prove this result very quickly, for if unboundedness is assumed then we can take for granted the existence of multiples as large as we wish. Hence the antecedent will at once imply

$$m \circ (n \circ a) \succ m \circ (p \circ b) \ \& \ n \circ (m \circ a) \approx n \circ (q \circ b)$$

and therefore

$$(m \cdot n) \circ a \succ (m \cdot p) \circ b \ \& \ (n \cdot m) \circ a \approx (n \cdot q) \circ b$$

whence

$$(n \cdot q) \circ b \succ (m \cdot p) \circ b$$

and so finally

$$n \cdot q > m \cdot p$$

as desired. On the other hand my proof of this theorem (given on p. 177–8) must distinguish cases, and roughly speaking in the discrete case the problem is transferred to a problem concerning natural numbers, where multiples are of course available, while in the divisible case we achieve the same effect by dividing instead of multiplying. Either method achieves very much the same results, except that I sometimes have to add existential conditions (as for example on pp. 182n., 231n). We may sum up by saying that the assumption of unboundedness is certainly convenient, and it is of course a standard assumption of classical Euclidean geometry, but on the other hand we can perfectly well do without it. This is just as well, since we now have good reason to suppose that it is in fact *false* for a number of paradigm extensive quantities, e.g. mass and volume. There is, then, no excuse for retaining the assumption as a condition on any extensive quantity, and that is why I abandon it.

What takes place is, as I say, the thesis that every quantity is either discrete or infinitely divisible. This in turn depends on the principle of least upper bounds, and so those who do not accept the latter principle for all extensive quantities are apparently forced into making the false assumption of unboundedness instead. Alternatively, they might simply postulate the dichotomy into discrete and divisible quantities axiomatically, as is at present their procedure with Archimedes' postulate, whereas in my version this is again a consequence of the principle of least upper bounds. My third main application of least upper bounds is in proving the theorem of 'the existence of the fourth proportional', but here they would proceed rather differently. The theorem as I stated it on p. 237 says that if a and b are instances of any quantity ϕ-ness whatever and if ψ-ness is a *continuous* and *unbounded* quantity with an instance y, then

$$(\exists x)(\alpha :_\varphi b = x :_\psi y).$$

Because I assume the principle of least upper bounds for all quantities, I defined 'continuous' to mean the same as 'infinitely divisible', but without this principle in the background we must distinguish the two. (For example, the rational numbers are infinitely divisible, but not continuous in the usual sense, just because bounded sets of rationals may well lack least upper bounds.) So let us now redefine 'continuous' to mean *both* infinitely divisible *and* satisfying the principle of least upper bounds. Then we shall be able to prove the existence of the fourth proportional, under the conditions stated, without having to assume that least upper bounds exist for *all* extensive quantities. This is the method these authors prefer.

Now the existence of the fourth proportional gives rise to an important embedding theorem. We select any arbitrary instance c of any desired quantity ϕ-ness and an arbitrary instance z of some *continuous* and *unbounded* quantity ψ-ness, and we have first

(i) $(\exists x)(a:c = x:z)$

and in addition

(ii) $a:c = x:z\ \&\ b:c = y:z \to (a \geqslant b \leftrightarrow x \geqslant y)$

(iii) $a:c = x:z\ \&\ b:c = y:z \to a:b = x:y:$

(iv) $a:c = x:z\ \&\ b:c = y:z\ \&\ E!a \oplus b \to a \oplus b:c = x \oplus y:z$

(v) $a:c = x:z\ \&\ E!n{\circ}a \to n{\circ}a:c = n{\circ}x:z.$

To see the purport of this theorem clearly, consider just the special case in which ψ-ness is also a *minimal* quantity in the sense of p. 246, i.e. where no distinct instances of ψ-ness are equal in ψ-ness. Then we can re-express this theorem in the language of functions as follows. There is a function which takes every instance of ϕ-ness to some instance of ψ-ness, namely the function f such that for all a

$$x = f(a) \leftrightarrow a:c = x:z.$$

The existence of the function is given by (i) above (and the assumption of minimality), while (ii)–(v) imply that it is an embedding function in the relevant respects, i.e. that

$$a \geqslant b \leftrightarrow f(a) \geqslant f(b)$$
$$a:b = f(a):f(b)$$
$$E!a \oplus b \to f(a \oplus b) = f(a) \oplus f(b)$$
$$E!n{\circ}a \to f(n{\circ}a) = n{\circ}(f(a)).$$

In particular, if we assume that the real numbers are somehow independently given, then we may take them as the instances of our continuous, unbounded and minimal quantity ψ-ness, and the theorem then states that the instances of any quantity can be embedded in this way in the real numbers. Viewing things from the usual contemporary perspective, this allows us to assign a real number to each instance of the quantity, with relations between real numbers mirroring the relations between the different instances, and so is just what is required for the quantity to be 'fundamentally measurable'. The point is that to obtain this result, which is what our authors aim at, we need no *general* assumption of least upper bounds, for it is only the real numbers themselves which need to satisfy the principle.

From my own perspective this strategy must be modified a little. I do not assume an antecedent knowledge of the real number system, but rather use the theory of proportion to construct that system, or anyway to construct an analogue in which the ratios play the part of the real numbers. Therefore the existence of least upper bounds for real numbers would in my treatment be replaced by the existence of least upper bounds for ratios, when ratios are themselves conceived as instances of an extensive quantity. However, since my ratios are, roughly speaking, obtained by abstraction from the general theory of extensive quantities, their properties depend upon what can be established in that theory. In fact we find that the ratios will satisfy the principle of least upper bounds provided that there is *some* quantity or other which satisfies this principle and is infinitely divisible. (Of course discrete quantities satisfy the principle automatically.) More precisely, we can show that the ratios will be continuous and unbounded and minimal provided that at least one quantity is continuous. But again this premise could be assumed—or rather proved (as on pp. 249–53)—without laying it down in the first place that least upper bounds must exist for *all* extensive quantities. So my own approach could equally well proceed without adopting this as a general condition on anything that is to count as an extensive quantity.

Although this might present itself as a more economical procedure, I nevertheless prefer the course taken in the text. I have avoided the assumption of unboundedness because it appears to be false for several paradigm extensive quantities, but I see no comparable motive for avoiding the assumption of least upper bounds. For example, it is extremely difficult to see how we could ever have empirical evidence for supposing that the degrees of some putative extensive quantity had the structure of the rational numbers, rather than that of the natural numbers (in the discrete case) or the real numbers (in the divisible case). If evidence does not guide us away from least upper bounds, considerations of theoretical simplicity will surely guide us towards them, as is obviously the case in geometry. Of course, it is not too difficult to imagine evidence tending to show that some of our *other* postulates are not all satisfied by some quantity, but then one simply has to conclude that the theory of proportion is not after all applicable to it, and hence that numbers cannot be 'fundamentally' applied to it in the way we have been discussing. (It may, perhaps, be measurable in some derivative manner.) However, the only plausible example I can see of a putative quantity which satisfies our other postulates but not that of least upper bounds is the rational numbers themselves, and *that* loss strikes me as unimportant.

Two other comments are perhaps worth making. It would be possible to retain the assumption of unboundedness by adopting the non-extensional interpretation of our postulates sketched on pp. 163–5. However, if we decide that a purely extensional logic is too great an asset to be lightly abandoned, then unboundedness must be rejected. Something must take its place if our theory is to get off the ground, and the thesis that every quantity is either discrete or infinitely divisible certainly presents itself as the most obvious alternative. This thesis is a consequence of the principle of least upper bounds, but without that principle it would appear to be lacking in justification. The same point may be made about Archimedes' postulate, which is again a consequence of least upper bounds, but not easy to justify in any other way. (It does *not* follow simply from the restriction to finite wholes, as we saw on pp. 132–4.) If, then, we do not include the principle of least upper bounds in our basic postulates, these other required theses will appear to be somewhat arbitrarily assumed. This point is made even clearer when we recall that in my version postulates (1′)–(6′) are not presented simply as a set of unargued postulates but are themselves derived from more fundamental axioms. In particular, the principle of least upper bounds follows (via the principle of dependent choices) from a similar principle concerning the existence of sums in the theory of part–whole relations, *viz.* principle P2 on p. 119. If we attempt to dispense with that general principle for part–whole relations and confine our attention to what I called the 'elementary' theory of part–whole relations (pp. 112–6), then there would be no prospect even of deducing Archimedes' postulate, let alone the dichotomy into discrete and divisible quantities. Nor does it seem possible to find a way of adding to the elementary theory of part–whole relations some principle to replace P2 that will be strong enough to yield Archimedes' postulate but not strong enough to yield least upper bounds quite generally.

A second consideration which weighs with me is a general prejudice in favour of second-order systems. If we drop the principle of least upper bounds from our characterization of extensive quantities, and put no comparable principle in its place, then the postulates will constitute a first-order theory. However, it is well known that first-order theories which permit infinite models cannot be categorical, not even in the somewhat extended sense in which my system is categorical (pp. 245–8). On the contrary, quite unintended and non-standard models will be available, and as a result the postulates have no claim to characterize the concept of an extensive quantity in any satisfactory way. So on quite general grounds one can say that the postulates will have to include at least one second-order principle if they are to do justice to our intuitive understanding of the structures involved. The principle of least upper bounds then presents itself as the obvious candidate. Of course I recognize that this line of thought will seem quite unconvincing to those who doubt the intelligibility of second-order quantification, but on that point I can here do no more than refer back to my somewhat extended discussion of the matter in several places in Volume 1.

4. Conclusions

WE may now review the main results of this investigation of the concept of a number, and comment on their philosophical interest. Once more I shall treat first of the natural numbers and afterwards of extensions to the concept of number, and I shall begin by saying something of what we may call the *uses* of natural numbers, or their *applications*, before I come to 'the numbers themselves'.

We have now considered a variety of uses, of which the main ones are the numerical quantifiers (both weak and strong), numerical powers of relations (or 'indices of operations'), numerical ordinals, and numerically definite comparisons. In the first three cases we showed how to construct the 'theory' of these uses, applying in each case a method of turning an intuitive introduction by means of recursive equivalences into a formal theory which had some claim to be considered as 'purely logical'. The essential features of the method are that (i) we begin by introducing some more general logical concept (e.g. that of a quantifier or an operation on relations) so that we may define the numerical concept we are aiming for as a special case of this, (ii) we furnish a definition of equality between all instances of this concept, so chosen that the numerical instances will be equal according to this definition if and only if they count as uses of the same number, (iii) from the definition of equality and the recursive equivalences we form the appropriate definition of succession and thereby (using the notion of the ancestral) the definition of the numerical concept we are after, and finally (iv) in order to establish the required existential theorem we effectively apply our numerical concepts to *themselves* in a way which very clearly requires the whole construction to be type-neutrally interpreted. In view of the several examples of this method already developed, I hope that this rather brief characterization of its main features will suffice, and I should now like

to comment further on two of these features, the definition of
equality and the type-neutrality.

In all my examples of this method the equality relation
chosen to correspond to sameness of number was equivalence.
In Volume 1, Chapter 5 (pp. 154–8) I argued that weak numerical quantifiers which are equivalent over the whole of their
range must correspond to the same number, and subsequently
in Chapter 1 of this volume it was equivalence that we took as
the equality relation between strong numerical quantifiers,
between numerical powers of relations, and between ordinals.
It is worth pointing out that this use of equivalence is not
forced on us by any general feature of the method but is
deliberately adopted for independent reasons. It is true that if
we are to remain within the standard framework of classical
logical systems there are really only two possible ways of defining the required equality—either equivalence or Leibnizian
identity—but either choice may be accommodated to the
general method. If we were to introduce further apparatus to
give us a wider choice—say modal operators to permit us to use
a necessary equivalence—the same would remain true. All that
needs to be assumed about our defined equality is that it is
formally an equivalence relation (i.e. transitive and symmetrical) and that it at least *implies* equivalence, and that is hardly
a controversial assumption.

Suppose for example that upon reconsideration of the theory
of strong numerical quantifiers we wished to revise the relevant
definition of equality and proposed to substitute for equivalence
some other relation, which I abbreviate to **E**. The definition of
succession is then altered to

$$Q_1 S Q_2 \quad \text{for} \quad Q_1 E Q_2{}'$$

the proper ancestral ✶**S** of this new relation **S** is introduced in
the standard way, the improper ancestral is defined

$$Q_1 \ast S Q_2 \quad \text{for} \quad Q_1 \ast S Q_2 \vee Q_1 E Q_2$$

and the strong numerical quantifiers are then redefined as
those quantifiers that bear the new relation ✶**S** to 0. Assuming
that the new relation **E** of equality implies, but is not implied
by, equivalence, the result is that the quantifiers now counted
as strong numerical quantifiers will be a proper subset of those

that used to count as strong numerical quantifiers according to the earlier definition. (For example it may be that 'there are exactly $2+2...$', which was counted as a strong numerical quantifier on the earlier definition, is not counted as one on the revised definition, since it does not bear **E** to 'there are exactly $0\ {''''}\ ....$'.) However, this will make no very noticeable difference to the formal proofs, since the relation **E** will coincide with equivalence when that is restricted to what now counts as a strong numerical quantifier, i.e. when the variables are numerical variables according to the new definition of a numerical variable.

This is quite easy to prove, for since **E** will be transitive and symmetrical it is easily seen from the definition of **S** that

$$n\mathbf{S}p\ \&\ m\mathbf{S}p \rightarrow n\mathbf{E}m$$

whence we shall deduce, taking into account the definition of the numerical variables and applying induction,

$$n*\mathbf{S}m' \lor n\mathbf{E}m \lor m*\mathbf{S}n'.$$

But by hypothesis we also have

$$n\mathbf{E}m \rightarrow n \equiv m$$

and hence

$$n\mathbf{S}m \rightarrow n \equiv m'$$

whence by induction (and retaining the earlier definition of $\geqslant$)

$$n*\mathbf{S}m \rightarrow n \geqslant m.$$

Further, we already have a proof of the crucial theorem

$$n \not\geqslant n'$$

and this proof is quite unaffected by changes in the definitions of other concepts not occurring in the theorem[1], though of course in order to establish the theorem it may be convenient to work with the old criterion of equality, or indeed something

---

[1] Strictly, the new **S** does occur in this theorem, since it is used in the definition of the numerical variable n. However, the only feature of this definition relevant to the proof is that it legitimizes induction for numerical variables, and this is rather a property of $*$ than of **S**.

apparently even weaker (as with $\approx$ on p. 41). However, from these theses it follows that

$$n\mathbf{E}m \leftrightarrow n \equiv m$$

for we have established the three implications

$$n*\mathbf{S}m' \rightarrow n \geqslant m'$$
$$n\mathbf{E}m \;\;\;\; \rightarrow n \equiv m$$
$$m*\mathbf{S}n' \rightarrow m \geqslant n'$$

and we are also given that their antecedents are jointly exhaustive and their consequents (in view of $n \ngeqslant n'$) mutually exclusive. Hence the converses hold also.

Therefore, tightening the definition of equality for strong numerical quantifiers, though it may add a few extra steps, will certainly not disturb the main lines of the proof of Peano's postulates. The same evidently holds for our construction of powers of relations, and of ordinals, and that is why I say that the general method does not require any particular definition of equality.

On the other hand the method *does* require essential use of the concept of type-neutrality in order to secure the existential theorem needed to take the place of an axiom of infinity. It is true that the form in which this existential theorem is officially required will depend upon what equality relation has been adopted. For example if this relation is taken as extensional equivalence then we shall need to prove that each numerical quantifier (or whatever) actually has some true applications, while if the relation is taken as *necessary* equivalence then we shall need to prove only that each numerical quantifier *possibly* has some true applications, and similarly in other cases. However, I see no method of *proving* that each numerical quantifier possibly has some true applications except by inferring it *a fortiori* from the premise that it actually has some true applications, and if this is correct then from the point of view of *proof* we are no better off with the one definition than we are with the other.

As I pointed out earlier (Volume 1, pp. 113–4), the complete type-neutrality which allows us to apply the numerical quantifiers to *themselves* is not strictly required for the purpose of deducing Peano's postulates. For example it would be adequate to suppose that they are significantly applicable to all (monadic) predicates in the orthodox hierarchy of types, whatever their level, for it is easily seen that there must be infinitely many of these. To assume only this much type-neutrality is doubtless a safer assumption, for it seems much less likely to lead to an unexpected contradiction than does the full type-neutrality I have been claiming. Philosophically, however, this restricted assumption would be rather hard to justify. The basis for my claim to type-neutrality is that the universal and existential quantifiers are recognizably the same in each type, and if this is granted there can surely be no objection to extending the same recognition also to certain numerical quantifiers, namely those based on a counting relation which is also type-neutral in that it is built up from universal quantifiers and truth-functional connectives, as is equivalence or Leibnizian identity. (Similarly, in the case of powers of relations we start from the observation that such familiar operations as taking the converse of a relation are recognizably the same whatever the type of the relation, and then argue that this recognition can be extended to numerical powers.) Now whatever grounds we have for recognizing the universal quantifier as type-neutral in this way surely continue to apply when that quantifier is used to quantify over the numerical quantifiers themselves, and that is why I cannot see the reasonableness of allowing a partial, but not a complete, type-neutrality to the numerical quantifiers. One might perhaps feel that this is giving an unnecessary hostage to fortune, for I must admit that the logic of type-neutrality has not been much investigated in this book[2] and it *may* turn out that what seem very natural assumptions lead to distinctly embarassing results, in which case my position would have to be that even the limited amount of type-neutrality that is actually needed for the proof of Peano's postulates is equally suspect from a philosophical point of view. However, it seems to me that this hostage ought to be given.

Suppose, then, that it turns out that there is something seriously amiss with the concept of type-neutrality that I have been using. What then happens to these constructions of

[2] I have tried to take the investigation further in my book *A Study of Type-Neutrality* (forthcoming).

numerical quantifiers, powers, ordinals, and so on? Well, one thing that is quite obvious is that our actual understanding of numerical quantifiers (etc.) does not in practice depend on the formal proofs I have offered, and the constructions cannot anyway claim to lay bare a process of reasoning that we actually go through when mastering these concepts. In practice we should no doubt go on believing that to say 'there are n . . .' does not come to the same thing as to say 'there are $n + 1$ . . .', even though we had no way of showing that these quantifiers had any true applications and even though we became convinced that neither did have any true applications, for they would still count as relevantly *different* quantifiers. Now one might explain this difference as a consequence of the fact that n and $n + 1$ are themselves different numbers, thus apparently presupposing that our grasp of numerical quantifiers is dependent on our grasp of the numbers themselves and so quite upsetting the direction of explanation that these constructions were designed to serve. The difficulty with this suggestion is that it invites the question 'How do you know that n and $n + 1$ are different numbers?', and it is not clear whether this question could be answered. I return to consider this point shortly. However, it would seem equally plausible to rely rather on the fact that it is quite *possible* that there should be n things of a certain sort (e.g. apples) but not $n + 1$ of them, for we can specify what situations would count as meriting this description even if no actual situation in fact does[3]. So if extensional equivalence were to fail us as a criterion for equality between numerical quantifiers, we could certainly adopt a stronger equality relation instead, and it might quite plausibly be said that a stronger relation is anyway a better approximation to our ordinary way of thinking. In my own opinion a necessary equivalence would in fact be too strong (Volume 1, p. 158), and the most plausible candidate for a strengthened relation which matches ordinary ways of thinking would be a non-material biconditional with the force of a counterfactual. However, I have no suggestions to make about the proper analysis of counterfactuals and so will continue to speak of the explicitly modal alternative.

---

[3] If we push the question 'how do we know that *this* is possible?', granted that it is not actual, there seems no answer that does not in the end appeal to our own 'clear and distinct ideas'.

One effect of thus dropping the type-neutrality would be that the theory of the numerical quantifiers (or ordinals, or powers) would lose any title it had to be a 'purely logical' theory for it would presumably have to embody an axiom corresponding to the axiom of infinity, stating that for each number n it is at least possible that there are n of something. However, as I have already indicated (Volume 1, pp. 199–205), I do not put much store by claims to be pure logic, because the crucial definitions of any such construction are themselves claims about how a familiar concept should be analysed, or what consequences flow from its analysis, and it seems futile to argue that these claims are themselves claims of pure logic. And *is* there any ground for saying that it is no part of the *analysis* of the concept 'there are n' that it is possible for this concept to have some true applications? On the other hand, if we set no value at all on the possibility of representing a construction as a 'purely logical' one, then the alternative construction of ordinals which I gave on pp. 45–8, resting on axioms for the notion of a sequence, might seem to be an equally acceptable construction of ordinals, and in this case no type-neutrality would be needed because the axioms for sequences themselves guarantee the existence of infinitely many sequences. Again we might begin by assuming that there are quantities satisfying the postulates for extensive quantities given on p. 150–1, and then introduce numerical variables for the first time in contexts of numerical comparison by applying the standard method to the recursive equivalences (p. 154)

$$(\forall x)(x \approx 1 \circ a \leftrightarrow x \approx a)$$

$$(\forall x)(x \approx n' \circ a \leftrightarrow x \approx n \circ a \oplus a).$$

However, it seems an evident objection to this last suggestion that one would not want the theory of numerical comparison to depend upon the premise that there are extensive quantities if that is a straightforwardly empirical assumption, and it would seem to be an empirical assumption if the postulates cannot be type-neutrally interpreted. The alternative assumptions about sequences are not straightforwardly empirical in the same way, since sequences are certainly intended to be *abstract* objects, but—as we have several times remarked (Volume 1, pp. 33–43, this volume, pp. 57, 147–7, 159–61)—our intuitions about which abstract objects there are should always be treated with some suspicion. One might perhaps infer that the mere assumption that something is *possible* (e.g. there being *n* of something, for each *n*) is always to be preferred to the assumption of the actual axistence of anything, abstract or not.

Since it seems to me that the type-neutrality assumption is in fact quite soundly based, it seems to me that this dilemma does not arise, but I thought that it deserved a brief discussion in view of the risk involved in proposing new and powerful logical systems. If the dilemma does arise it will evidently be of some importance if we are intending to go on to construct arithmetic proper, i.e. the theory of the numbers themselves, from (one or other of) our constructions of these uses of numbers. One would certainly wish to avoid a situation in which the truths of arithmetic were seen to depend on an evidently empirical assumption, and I would equally wish to avoid resting them on assumptions about abstract objects such as sequences, or indeed sets. The only abstract objects which it might be reasonable to assume, when constructing the elementary theory of numbers, are the numbers themselves. But do we have to posit nuumbers as objects? Or can we regard numbers as analysable without remainder in terms of what we are calling their uses, i.e. the numerical quantifiers, powers, ordinals, and so on? Here we once more have to face what is now a very familiar objection.

It is an objection to the proposal that natural numbers be identified with sets that any number of different identifications may be suggested, all with an equal claim to correctness. For since they cannot all be correct we must conclude that none are. This same objection was also raised to the proposal that rational numbers be identified with pairs or sets of pairs of natural numbers and to the proposal that real numbers be identified with cuts or sequences of rational numbers. The objection does not depend upon the fact that in each case the usual proposal is to fix upon some already known *objects* to identify with numbers, for it remains even if we reform the proposals so that the statements of number theory are to be translated *as wholes* without any reference being assigned to the numerals occurring within them. It is still true that if different translations are available, each with an equal claim to correctness, then either all are correct or none are. This objection appears to destroy the suggestion that the statements of arithmetic be analysed as statements concerning numerical quantifiers, or powers of relations, or ordinals, or anything else of that sort. To put the matter crudely: natural numbers have *many* uses, and therefore it cannot be right to identify them with any one of their uses.

The only escape from this objection would seem to be to maintain either that some one of the uses of natural numbers is more fundamental than any of the others, or that each of the different translations we have suggested in fact has the same meaning, so all are equally correct. The latter seems to me a most implausible position. For example, we have suggested rendering '4 is the next natural number after 3' as a statement involving weak numerical quantifiers, roughly

There are at least 4 so-and-so's if and only if there is some so-and-so such that there are at least 3 so-and-so's other than it

and similarly as a statement involving strong numerical quantifiers, roughly

there are exactly 4 so-and-so's if and only if there is some so-and-so such that there are exactly 3 so-and-so's other than it

and again as a statement involving powers of relations which seems to have no very natural English equivalent except

one thing bears the 4th power of a relation to another if and only if it bears the relation to something that bears the 3rd power of the relation to that other

and we could also mention separately the interpretation in terms of 'indices of operations'

operating 4 times is the same as operating once more on the result of operating 3 times

and finally there is the translation involving ordinals, roughly

the 4th so-and-so in some order is the next so-and-so in that order after the 3rd so-and-so in that order.

Of course further variations can be suggested. Since there are no generally agreed criteria of when two expressions have the same meaning, there is no evident way of demonstrating that these 'translations' do not all have the same meaning as one another, but I take it that the point is not likely to be controversial. If argument is needed, it may be relevant to remark one curious point of difference. On the suggested schemes of translation involving quantifiers or ordinals, if the relevant generalizations are taken not to be type-neutral but restricted to objects, then in order to prove that 3 and 4 are distinct numbers it would be necessary to prove that there are at least three distinct objects. However, if the scheme of translation involving powers of relations (or indices of operations) is taken in the same way, then we only need two objects to prove that n and $n+1$ are always distinct numbers. Of course with the translations all type-neutrally interpreted we shall not find any arithmetical statement which is assigned one truth-value on one scheme and a different one on another. But such divergences do arise if the schemes are all modified in the same way, by dropping the type-neutrality. This seems quite a strong ground for saying that even with the type-neutrality there must still remain a difference of meaning.

Nor does there seem much prospect of success for the alternative defence that one of the translations is somehow more fundamental than any of the others. As we have seen, each is as good as any other for the purpose of constructing arithmetic, in that each of these uses of number can be introduced independently, without relying on any previous understanding of number in any form, and in each case the relevant theorems can be proved by very much the same method. The availability of independent introductions confirms, what is surely clear to intuition anyway, that it would be entirely possible for a man to master any one of these uses without yet having encountered any of the others. Therefore no priority can be assigned on this score. On the other hand any of the uses can be adequately introduced by being defined in terms of any of the others, so none of them *has* to be given an independent introduction. It is true that in some cases the definitions are very simple to give, and in others they are more roundabout. For example it is easy to define ordinals in terms of strong quantifiers but not quite so straightforward to reverse the procedure, and it is comparatively simple to define weak quantifiers in terms of powers of relations but the reverse definition strikes one as noticeably artificial. Of the uses which we have considered, the one in terms of which it seems easiest to define all the rest is that of ordinals applied to the terms of sequences, but this has the drawback that it presupposes our axioms for sequences which surely do not have to be grasped in order to understand, say, numerical quantifiers. If we are not to bring in such extra axioms then probably the use which has the best claim on this score is the numerical powers of relations, which is interesting because this use seems furthest removed from ordinary English idiom. Presumably such trifling considerations as these cannot establish that one of our translations has a better claim to correctness than the others, and the objection is not to be defeated this way.

If we cannot fairly pick any one of our uses of numbers as what yields the concept of the numbers themselves, must we conclude that our grasp of the numbers and their properties is in fact independent of our grasp of all their uses? This would surely be an error. To take an analogy, it may perhaps be pointed out (say to readers of Plato's *Republic*) that the adjective 'just' has a number of distinct but connected senses. For example we speak of just men, of just actions, of just laws, and (at any rate according to Plato) of just states. Now it does not seem very plausible to maintain that the sense of the word 'just' is exactly the same in each of these contexts: one would not expect a paraphrase that fits one of these contexts to be transferable unchanged to the others (a point which Aristotle noticed about some other adjectives, e.g. 'healthy man' and 'healthy climate'). On the other hand the senses are evidently connected, and one *might* find—though I suspect that the analogy halts a little here—that no one sense is central but rather that any of the senses can be defined in terms of any of the others, and any of the senses can be mastered before encountering any of the others. Supposing this were so, it surely would not follow that we must abandon the claim that our concept of justice and its properties is dependent on our grasp of the uses or applications of that concept as displayed in our mastery of the adjective 'just'. However, the situation would certainly provoke problems for the traditional reductionist way of exhibiting this dependence, *viz.* by producing a scheme for translating any sentence containing the word 'justice' into a corresponding sentence containing the adjective 'just' instead.

As a matter of fact many occurrences of the noun 'justice' could quite reasonably be said to concern only one sense of the adjective 'just'. For example when it is said that justice is a virtue no doubt it is only just *men* that are in question, since virtues are human dispositions; on the other hand a phrase such as 'rough justice' evidently relates to *acts* that are (in some 'rough' way) just, while trivially if the justice of some law is in question what is in question is whether that law is a just *law*. Therefore if we imagine, to create a closer parallel with the verbal distinction between 'two', 'twice', and 'second', that each different sense of the adjective 'just' is represented by a different word, then we shall find that we wish to paraphrase some uses of 'justice' with one of these words and some with another. This is not, however, a difficulty of principle. What is more awkward is that some occurrences of 'justice' seem to concern *all* uses of the adjective 'just', as for example in 'justice is always desirable', so here the reductive paraphrase would apparently have to be a conjunction with one conjunct for each different sense of the adjective. (Remember that we are assuming, *argumenti causa*, that no one sense of the adjective is central.) The adequacy of the paraphrase would then require that we should have at hand a list of all such senses and a reason for supposing that they were all. But perhaps in the case of justice we could with some plausibility maintain that statements calling for such a conjunctive analysis occur only seldom, and when they do they tend to be so vague that it might be better to treat them simply as ambiguous.

This plea will not work for the numbers. As we have seen, the very elementary arithmetical truths, such as $7 + 5 = 12$, can be quite directly translated into all the uses of numbers we have considered, and since *any* arithmetical statement *can* be defined in terms of identity and succession, and these have direct translations into all uses, there is no case in which we have to say that only some uses are relevant. It appears, then, that the traditional style of reductive analysis must in this case consist of a conjunction with one conjunct for each use, and it is very doubtful that the condition of adequacy just mentioned could be met. More importantly, however, the whole style of analysis is in this case quite implausible, for although no one of the uses of numbers has a good claim to be central, equally it does not appear that *all* are essential, and it can hardly be right to imply that a man who knows nothing of, for example, numerical powers of relations is thereby shown to have only an imperfect grasp of the sense of $7 + 5 = 12$. On this ground, then, the traditional style of reductive analysis does not appear to work for numbers.

Although this objection is I think the most interesting, it may be relevant to consider another reason that has been advanced for saying that a reductive analysis ought not to be applied to numbers, even though it may be perfectly legitimate elsewhere. It is clearly a feature of our common ways of talking that we have no hesitation in employing what is, from a grammatical point of view, a quite unrestricted abstraction principle. We happily form abstract nouns or noun phrases in a variety of ways without thinking of ourselves as introducing any interesting existential assumptions thereby, and where for the most part we readily agree that the abstract nouns are introduced only for brevity (as for example with this last occurrence of 'brevity'). Indeed one of the commonest types of obscurity in speech and writing is the tendency to use abstract nouns in such a way that we cannot see what longer and more concrete locutions they are supposed to be replacing, and when in elucidation the abstract nouns are paraphrased away we certainly do not think that the resulting gain in clarity has been achieved only at the cost of a distortion of the ontology. So it is quite reasonable to hold that the technique of eliminating abstractions to increase clarity is not only of literary value but also of philosophical relevance; the mere use of the grammatical noun form is not of itself of any noticeable significance and it need not be suspected that the reductive paraphrase must always leave out something important.

Now this thesis, that the occurrence of a noun form should not be taken too seriously, is particularly persuasive when the noun in question only occurs in a rather limited range of contexts, as with Quine's example 'sake' (Volume 1, pp. 44–5), and Dummett has proposed that differences in this respect should be regarded as critical[4]. At any rate if we base ourselves

---

[4] Dummett (1973) pp. 70–3, 78–9. In between Dummett associates this test with the existence of a special 'criterion of identity' for the objects in question, but as he admits (p. 79) that test is often rather inconclusive.

upon the Fregean notions of object and concept we shall observe
that one of Frege's characteristic doctrines was that *every*
concept should be capable of taking *every* object as argument,
which is no doubt rather an idealization of actual usage but one
that has at least some basis in that part of our language which
concerns concrete objects. (It is also an idealization firmly built
into the classical predicate logic, in so far as it is stipulated
that *any* predicate-symbol of degree n, when followed by *any* n
terms, must yield a well-formed formula.) The suggestion is,
then, that the ability to occur in a wide, if not all-embracing,
range of contexts is from a philosophical point of view crucial
to being a genuine name standing for an object. However,
Dummett then urges that most abstract nouns, like 'justice', in
fact occur only in a rather small range of contexts, and this
is one of the things that distinguishes them from names of
numbers, since we have a very well developed vocabulary for
talking of numbers. Numbers, then, ought to be admitted as
objects, though virtues (for example) are denied that status.

Now I confess that, as stated, this argument seems to me
very weak indeed. The Fregean scheme requires that there be
just one category of names and just one category of (first-level)
predicates, and that any combination of name and predicate
yields a significant sentence. However, the situation in ordinary language is *more* like that required by a Russellian theory
of types, for although there are indeed plenty of predicates
applicable to numbers those predicates are not usually applied
to concrete objects, and *vice versa*. No matter how extensive
our vocabulary for speaking of numbers, if that vocabulary is
in fact isolated from the vocabulary we use when speaking of
what are uncontroversially objects, the mere extent of the
vocabulary hardly seems a reason for classifying numbers as
objects.

Perhaps, however, we can extract a slightly different argument from what is essentially the same starting point, for *why* is
it that our language contains many more contexts in which
numerals occur as nouns than it does contexts in which, say,
the noun 'justice' occurs? Presumably the answer is that numbers fall into a system which gives them interesting relations
to one another, whereas the relations between justice and, for
example, the other virtues are altogether more simple and ele-

mentary. As a result we seldom have occasion to make statements about justice which do not have pretty direct relevance
to the application of that concept, and for which therefore a
paraphrase using the adjective 'just' comes to mind readily
enough. With numbers, on the other hand, the ways in which
they are related to one another come to be of independent
interest, just because of their complexity, and we therefore
introduce a special vocabulary for speaking of this. For
example we have observed that addition has a simple and
direct translation into all the uses of numbers we have considered, and the same can perhaps be said of multiplication.
But what of exponentiation? The terminology of 'squared'
and 'cubed' points of course to a geometrical application, but
this is relevant only for powers up to 3, and beyond that it is
probably fair to say that there is no reasonably *simple* translation of exponentiation into any of the uses we have considered.
(Of course there is always a longwinded one.) We learn to understand, and use, arithmetical statements such as $2^4 = 4^2$ without
having any particular application in mind. The point can perhaps
be made more strongly. A man who used the word 'justice' but
was quite unable to paraphrase his remarks in terms of the
adjective 'just' (or an equivalent), and could not see the relevance of such a paraphrase when offered, would be regarded as
having no grasp of the concept of justice. However, this is surely
not our reaction to one who uses the terminology of exponentiation but cannot see how to paraphrase his remarks in terms of
numerical quantifiers, or ordinals, or whatever, and who does not
regard any such complicated paraphrase as relevant to his
meaning. Indeed it is perfectly clear that once we progress beyond very elementary truths of arithmetic we do not in fact regard ourselves as using an idiom of abstraction which is capable
of paraphrase in a clearer but more longwinded form, and to
most of us the possibility of a lengthy reductive analysis no
doubt comes as a complete surprise.

Finally I should perhaps mention one further objection which
is in effect an objection to all such reductive analyses. If all
intelligible remarks about natural numbers can be paraphrased
into remarks containing numerical quantifiers (or ordinals,
etc.), then 'Julius Caesar is not a number' is not an intelligible
remark, though it certainly seems to be. Similarly, if all

intelligible remarks about the virtues can be paraphrased into remarks containing the adjective 'virtuous', then 'Julius Caesar is not a virtue' is again unintelligible. Once more, if all intelligible remarks containing the noun 'sake' can be paraphrased in such a way that that noun occurs only in the context 'for the sake of', then 'Julius Caesar is not a sake' is not intelligible. With the last example one might agree that it is not intelligible (though even here a doubt seems possible), but with the first two the contention has no initial plausibility. The fact is that although our vocabulary for speaking of numbers is in practice pretty well isolated from our vocabulary for speaking of concrete objects, there are certainly exceptions, and we can trade upon the similarity of grammatical form to link the two. If he is not to brand these remarks as unintelligible, the best that the reductionist can do with them would seem to be to reconstrue them as metalinguisitic, for example as a misleading way of saying that 'Julius Caesar' is not a numeral. The only alternative seems to be to admit that they concern an intelligible, but mistaken, theory according to which numbers are objects (cf. pp. 165), But if the theory is intelligible what is the reason for supposing that it is mistaken?

Now the first of these three objections, that there is no way of choosing which reductivist analysis to adopt, does not overthrow the general contention that our grasp of the numbers themselves depends upon our grasp of their uses, but it does apparently show that the very straightforward account of this dependence which is embodied in a traditional reductive analysis is too simple. The second objection, though I consider it rather inconclusive by itself, may again be taken to suggest that traditional reduction is oversimplified. For one could regard it as granting for the sake of argument that at a very elementary level, as when children are taught to add by playing with beads or blocks, the reductive analysis may apply, but then pointing out that mathematics develops beyond this very elementary level. As for the last, very general, objection, all I can say is that it seems to me to be quite unpersuasive, though I do not know what, in general, is the best way to meet it. At any rate in the case of numbers a solution will emerge from my treatment of the first two objections.

What are the alternatives to a reductivist account of the

numbers? Well, one traditional rival is the Platonic conception of numbers as objects existing in their own right—non-concrete objects of course, but nevertheless in some way open to our mental gaze. We do know some truths about them, but this is in no way connected with our ability to use numerical quantifiers, ordinal adjectives, and the like; and the way that we find out that, say, $3 \neq 4$ is simply by inspection of these very objects, the number 3 and the number 4 themselves. This picture is surely unintelligible, for the alleged method by which we find out about the numbers is completely mysterious. In particular it seems to leave open the possibility that we have all been undergoing some 'mental hallucination' about the numbers, and in fact $3 = 4$, though this will *not* imply that for there to be 3 so-and-so's is the same as for there to be 4 of them, the 3rd item in some order need not on that account also be the 4th, to do something 3 times may still be distinguished from doing it 4 times, and so on. The latter non-identities involve concepts which we can clearly master independently of any sort of perception of certain special objects called the numbers, and on this view there is no reason to suppose any sort of connection between our knowledge of the one and our knowledge of the other. Nor is the situation much altered if the numbers themselves are moved from their Platonic heaven and relocated in the mind as 'mental mathematical constructions', though the metaphor of 'mental perception' may perhaps seem easier to swallow in this case. For both Intuitionists and Platonists appear to believe that our knowledge of the numbers themselves is completely independent of our mastery of those concepts, such as numerical quantifiers, which I have been saying embody *uses* or *applications* of numbers, and consequently it could perfectly well turn out that something had gone wrong with our grasp of the one while leaving our grasp of the other quite unaffected. But this is an absurd position[5].

We began then with some arguments designed to show, to put it crudely, that the numbers themselves should not be identified with their uses, and we have now briefly argued that

---

[5] Whether there are any Platonists in this apparently exaggerated sense of the term may perhaps be doubted, but—apart from Plato himself—Gödel is quite a good candidate (see pp. 294–5).

they should not be divorced from their uses either. What intermediate position remains? Perhaps we could picture the situation in this way. We begin by mastering, say, the concept of a
numerical quantifier, so that we can use these quantifiers
effectively (for example in counting) and perhaps perform with
them some simple operations such as addition. But then we
also catch on to the idea of an ordinal, or perhaps a numerical
index of operations, or a numerically definite comparison, or
something else of this sort. It can hardly fail to strike us that
there is a structural analogy between these systems of concepts, and that what can be discovered to hold for one of them
can be automatically transferred to the others. (However, if
language is any guide, this may not strike us absolutely at once;
for example there is no common root to *duo*, *bis*, and *secundus*.)
Then perhaps we can think of the next step as arising from a
demand for explanation—why do these analogies hold?—or
just from a demand for simplification—let us have *one* theory
which systematizes *all* such facts as that two of one thing and
three of another makes five in all, doing something twice and
then doing it three times more is doing it five times altogether,
taking the second term of an order and then going on to the
third after that results in the fifth term, and so on. The result
is that we then adopt something like the standard formal
theory of arithmetic, though no doubt in an imperfectly axiomatized form, in which we speak as though we were concerned
with certain objects called numbers. This step could be taken
just as a way of simplifying the vocabulary of some one use of
numbers, perhaps to facilitate further investigation of the
structure, and in that case a traditional reductive analysis
would seem applicable. (It could be argued that the analysis
would still be applicable even though in practice we lost sight
of applications when engaged in our arithmetical investigations.) However, if the step is taken as a unifying move, to
enable us to treat at once of several uses, this reductive analysis
is no longer available, and perhaps a better approach is to
think of the arithmetical theory on the model of a scientific
theory. The data which give rise to the theory will therefore
consist of facts about numerical quantifiers, ordinals, and so
on, that we can grasp quite independently of any such theory,
and we need not stickle over just which facts about the uses

of numbers may be regarded as evidence preceding the formulation of the theory and which are subsequent deductions and applications. The main point is that this is where the evidence for the theory lies, but the theory is not just a restatement of the evidence. Its point is to systematize, unify, and perhaps explain the evidence. It may also furnish predictions, i.e. new facts about numerical quantifiers and the like which we had not originally been aware of.

Since ordinary scientific theories may themselves be conceived in a number of ways, this basic approach still leaves plenty of room for manoeuvre. In particular we may distinguish the *instrumentalist* and the *realist* approach to theories, which differ primarily over the attitude they take to the 'theoretical entities' introduced by the theory, namely in our case the numbers. According to the instrumentalist approach the fact that the theory will be presented in a language which apparently speaks of a new kind of object is not to be taken seriously; it is not to be supposed that the world actually contains any such objects, for the task which the theory has to perform does not in any way require their existence. To this the realist will usually object on two counts. First he will complain that the instrumentalist has too narrow a view of what the purpose of a theory is. If its only job was to serve as an instrument for deriving predictions, then indeed it would not matter too much how it did so, so long as it was successful in this task. However, we also look to scientific theories not just to tell us what will happen but also to explain how and why it happens, and for this purpose the theory must be taken as an attempt to give a genuine description of what actually goes on. Besides, the theory's success in yielding predictions would itself be mysterious unless there was in fact some connection between the way the theory worked and the way the world worked. A second objection could be made on pragmatic grounds, for the history of science seems to show that it is by taking theories realistically that science progresses; it is by thinking of the theory as a genuine description of what goes on that we are able to suggest ways of extending the theory and to see where it might require modification. Thus, to use a stock illustration, if the kinetic theory of gases is taken realistically this will suggest questions about the number of molecules in a given mass of gas (Avogadro's hypothesis) and will also suggest possible corrections which take account of the properties which a genuine molecule might be expected to possess, such as size (Van der Waals' equation). If the term 'molecule' is taken only as a meaningless cog in a machine for deriving predictions, it is impossible even to understand these suggestions. Generally, then, in scientific matters it will be argued that a realistic approach to theories is required to satisfy our demand for explanation and has in fact proved fruitful. However, such arguments do not apply with equal force to all scientific theories —see, for example, Berkeley's *De Motu*—and it is not clear that they apply at all to the theory of numbers.

On the assumption that my constructions of numerical quantifiers, ordinals, and so on are satisfactory, we certainly cannot expect that a realistic approach to number theory will be any more fruitful than an instrumentalist one, for as we have seen the theory of natural numbers can in fact be reduced without loss to any of these theories of the uses of number, and therefore anything we might discover by using ordinary arithmetic could equally have been discovered without introducing the numbers themselves at all. One might perhaps say that we find it simpler to operate with the conception of numbers as objects, and this will therefore facilitate discoveries, but that of course is entirely consistent with the claim that numbers are just useful theoretical tools and it need not be supposed that there actually are such things. Therefore I cannot see how a realistic construal of arithmetic can be expected to increase our knowledge of the numbers and their uses.

Nor does it seem to have any extra explanatory power. Gödel has argued for Platonism in mathematics on the ground that

> 'the assumption of such [abstract] objects is quite as legitimate as the assumption of physical bodies and there is quite as much reason to believe in their existence. They are in the same sense necessary to obtain a satisfactory system of mathematics as physical bodies are necessary for a satisfactory theory of our sense perceptions.'[6]

The suggested analogy is that just as the failure of the phenomenalist reduction of physical objects to sense data shows that

[6] Gödel (1944), p. 220.

we cannot avoid postulating physical objects as part of a theory to explain why we obtain the sense experiences we do, so equally the failure of the nominalist reductions of abstract objects shows that we cannot avoid postulating abstract objects as part of a theory to explain why we obtain the 'mathematical experiences' we do. Since Gödel was at this point presupposing that any satisfactory nominalist reduction would have to be predicative in character, and my suggested reductions are certainly not predicative, it may be rash to suppose that he would apply the same argument to the present problem. However, he does apparently think that there is something which can only be explained by postulating abstract objects, namely our 'mathematical experiences'. As for what these mathematical experiences are supposed to be, he tells us elsewhere (speaking of sets) that

> 'despite their remoteness from sense experience, we do have something like a perception also of the objects of set theory, as is seen from the fact that the axioms force themselves upon us as being true. I don't see any reason why we should have less confidence in this kind of perception, i.e. in mathematical intuition, than in sense perception.'[7]

Applying this view to the natural numbers, the argument is that one must take seriously the existence of numbers as abstract objects in order to explain the fact that the axioms of number theory 'force themselves upon us as being true'. However, it is not the least bit clear how the actual existence of numbers would in any way help to explain this fact, and an alternative explanation is evidently provided by my suggestion, since the axioms of number theory are designed to mirror the truths about numerical quantifiers and the like, and these latter 'force themselves upon us as being true' in just the way that other logical truths do. The more usual argument for Platonism in mathematics is not that it is needed to explain our intuition of mathematical truths, but that it is needed to explain the success of applications of mathematics in science and in daily life. However, this argument is of no force in the present context, for the whole point is that the various *applications* of arithmetic can be independently accounted for. Nor,

---

[7] Gödel (1947) p. 271.

finally, do I see that the structural similarity between the systems of numerical quantifiers, relational powers, ordinals, and so on is in any way *explained* by postulating a set of objects whose *raison d'être* is to exemplify that structure in yet another form. So far as I can see, then, there is no advantage to be gained from a realistic construal of the theory of numbers, and I would suggest that it is best taken merely instrumentally; its purpose is to systematize and unify, which it does perfectly well, and its apparent postulation of objects called numbers is not to be taken seriously.

This is in a sense a formalist position, in so far as the formulae of arithmetic are denied a literal truth-value and their usefulness is explained in terms of their applications. Where I intend it to differ from the usual expositions of pure formalism is in the account of how the formulae and their applications are related. According to the usual formalism, we begin with a set of uninterpreted axioms and rules, and we then find that the resulting set of derivable uninterpreted formulae can be given some useful applications. The only reason that can be advanced for beginning with one set of axioms and rules rather than another is that the resulting system is aesthetically satisfying and proves to have more useful applications than its rivals. As for the nature of these applications, this is a matter on which formalists tend not to be forthcoming, for, *qua* pure mathematicians, they consider this to be not really their concern. But perhaps we may take as typical this passage by Curry:

> 'Perhaps this is the place to deal with the rather silly objection which Ramsey advanced against formalism of the Hilbert type —*viz.* that in saying "it is two miles to the station", "two" must have a meaning which the formalist gives no account of. But it is easy to supply such a meaning. In fact it was, in its essentials, done by Dedekind; a class has n members if it is similar to his Z_n. Of course it was never intended that a formal system should be completely isolated from its applications.'[8]

The general line of thought here is that the numerical quantifiers are to be explained in terms of one–one correspondences with certain sets of symbols in the uninterpreted system— thus there are two miles to the station if and only if the set of

[8] Curry 1951, p. 67.

miles to the station is in one–one correspondence with the set of uninterpreted symbols {'0', '1'}—so apparently one would not understand what was meant by 'there are two miles to the station' unless one was already familiar with the uninterpreted system. On my account of the matter the direction of explanation is entirely reversed. What are called applications come first, and these may be understood and their properties established quite independently of any uninterpreted system. Further, the system of arithmetic is designed precisely to mirror the properties of these 'applications', and a suggested arithmetical formula is acceptable or unacceptable for inclusion in the system according as it does or does not mirror a truth concerning numerical quantifiers, indices, ordinals, and so on. Thus we can explain, what must be mysterious on the usual formalist account, how number theory could be a flourishing subject long before anyone had even thought of axiomatizing it, and we can explain our conviction that it is not up to us to choose whichever axioms we please. Since the 'applications' have perfectly good truth-values and since an arithmetical formula counts as acceptable if and only if the corresponding 'applications' are true, it is entirely natural to speak of the arithmetical formulae themselves as true or false. If we do, however, we should recognize that the truth in question is not quite the ordinary 'correspondence with reality', for any specification of just which features of 'reality' the formula is to correspond to is bound to be somewhat arbitrary.

When we turn to consider the way in which the concept of number has been extended to include negative, fractional, irrational, and imaginary numbers, I think we find that the same type of account is still applicable and for much the same reasons. Certainly no extra reason emerges for taking numbers seriously as objects. Historically these extensions came about in rather different ways, but it is broadly true to say that negative and imaginary numbers were first introduced without any particular application in view but simply in order to provide missing solutions to mathematical equations, whereas rational and real numbers were in the first place called for by their applications. (It is interesting that in the sixteenth and seventeenth centuries there were quite a number of mathematicians who rejected negative numbers as fictional and unreal entities, and in the eighteenth and early nineteenth century the same was true of imaginary numbers, as the name 'imaginary' itself shows[9]. On the other hand there seems to have been no such reluctance to extend the Greek concept of number to include both rational and irrational numbers.) I would judge that most people still think of imaginary numbers in this way, though perhaps one day they will seem just as central to the study of plane geometry as Cartesian co-ordinates do now. On the other hand we have no lack of applications for signed numbers—for example the association of + and − with opposite directions is very natural and easy—and it would be a relatively simple task to give a construction of signed numbers which took (one of) their applications as basic and which was at least as satisfactory as my construction of rational numbers from their applications. In the other direction, although it is clear that as a matter of history the impetus and guide to the study of real numbers was their geometrical application, one might say that this is no longer the way real numbers are regarded. As a result of the 'arithmetization' of analysis in the last century, in many contemporary authors 'analysis' has pretty well come to *mean* the theory of sets (or sequences) of natural numbers.

The appropriate moral to draw from these rather cursory historical remarks seems again to be that there is no *one* way of introducing these various extensions to the concept of number which has a satisfatory claim to be the right one. Thus it is certainly plausible to claim that a man who knows that $\frac{3}{4}$ is supposed to be a number which results when 3 is divided by 4 but who does not know what it is to be three-quarters of something is lacking an important ingredient of our concept. However, the converse claim is equally plausible, for it is certainly part of our concept that $\frac{3}{4}$ is a number like any other number, capable of being added, multiplied, divided, and so forth, and less than 1 in the same sense as that in which 1 is less than 2. So it would seem that a proper introduction of rational numbers must *both* make clear their relation to natural numbers *and* indicate their use as fractions and as ratios. But again, how far is the list of uses to extend? For example if we had presented the rational numbers by speaking only of

[9] See for example Kline 1972, pp. 251–4, 592–7.

their use as fractions, this could reasonably be said to be an incomplete explanation because in our theory $\frac{3}{4}$ and $\frac{6}{8}$ are counted as the *same* number though there is quite reasonable ground for regarding them as different, though in a certain sense equivalent, fractions. However, if we then add to this approach the approach that sees rational numbers as required for the purely arithmetical role of ensuring that division is always performable, we could see how the needs of the latter could be taken as prevailing and we could see too how this might lead us to extend the concept of a fraction to include improper fractions. So far, however, we have had no call to mention ratios or the use of rational numbers in simple statements of numerical comparison, and perhaps this could be regarded as a rather optional extra. On the other hand we might have passed over the ordinary concept of a fraction altogether, i.e. of a *part* of something related in a certain way to the whole, and in effect gone straight to the notion of a ratio without mentioning parts at all. As I argued in Chapter 2, §2, if we do entirely omit the notion of a part and simply start with the traditional postulates for extensive quantities, we shall not have an adequate account of *our* concept of a numerical comparison. However, it could be replied that we should have quite enough to understand the interest that attaches to the formal postulates for rational numbers, and quite sufficient 'evidence' for the 'correctness' of those postulates. To take a more extreme example, the notion of a '3 in 4' chance is doubtless fairly peripheral from our point of view, but it is presumably conceivable that there should be a race of men who used rational numbers *only* in contexts of probability, and though their concept could indeed be said to differ from ours we might still count it as the concept of a rational number if their rational numbers were related in the appropriate way to other varieties of number[10].

More generally, the orthodox logicist account, by which rational numbers are seen as logical constructions from natural numbers, seems to be just wrong because there are features of the formal theory which it leaves unexplained and because there is no account of the interest which might attach to this way of rewriting parts of the theory of natural numbers. If

[10] We may imagine that their theory includes rational numbers greater than 1, but that such numbers have no role outside pure arithmetic.

logical constructions are to be used at all, I prefer constructions which are based on the uses of rational numbers and are in this way parallel to the logicist constructions of natural numbers, because this approach certainly explains the interest attaching to rational numbers. But again we have too many to choose from, and again we shall not always be able to account for *all* features of the formal theory, in particular for the fact that rational numbers and natural numbers are seen as combining to form one coherent system. For example, one could object to my construction of rational numbers at the end of Chapter 2 that it does not satisfy the requirement that $\frac{3}{4}$ is to be less than 1 in exactly the *same* sense as that in which 1 is less than 2. We shall indeed deduce (p. 190)

$$\tfrac{3}{4} <_r 1 \ \& \ 1 <_r 2$$

but we shall also have (without subscript)

$$1 < 2$$

while

$$\tfrac{3}{4} < 1$$

will have to be undefined (provided we retain orthodox requirements on correct definitions). In order to obtain a clearly correct account of the relation between natural and rational numbers, as *we* conceive them, the formalist postulation of certain *extra* numbers to complete the original system in a certain respect seems to be required, and of course the new numbers thus postulated will have the same ontological status as the old. Yet this approach will not do by itself, because it cannot explain how we can *use* fractions and ratios, and know many of their properties, without even having thought of how to divide natural numbers.

Finally, one might wonder how important it is to reproduce *our* conception of the relationship between the different varieties of number, and here let us change the example to real numbers. Imagine, for example, that the classical Greeks had extended their theory of proportion on the lines sketched in Chapter 3, clearly treating ratios as objects capable of addition and multiplication and other such operations just like our numbers, but sharply distinguishing them from the natural numbers and seeing only a very superficial resemblance between, say, the ratio which anything has to itself and the natural number 1. We could imagine further that by reflection on the structure of the ratios they developed a differential and integral calculus entirely similar to ours and also came to make use of their system of ratios just as we now make use of our real numbers, for example assigning particular ratios to objects as measures of their length, mass, and so on, as on p. 243. Would it not be somewhat arbitrary to conclude that their concept of a ratio did not count as the concept of a real number, just because their ratios are not related to natural numbers in the way that our real numbers are?

One consequence would be that their concept of a 'real number' could be quite adequately handled by the first and simpler reductive analysis of Chapter 3, whereas ours cannot. It is significant that the difficulty we encountered in Chapter 3, §3 was just the difficulty of extracting from the system of ratios a system of real numbers appropriately related to natural and to rational numbers. We attempted to overcome this difficulty by introducing real number variables for the first time in contexts such as $a \approx_N b$, observing that propositions of this form could certainly be regarded as propositions about the ratio of a to b, but that at the same time this form could be appropriately compared with the forms $a \approx_n b$ and $a \approx_{(n/m)} b$. However, it was difficult to characterize these propositions for two reasons: first because it was not altogether easy to see how to pick out the propositions which one could regard as assigning a *numerical* value to this ratio, and so fixing its position with respect to the ratios between natural numbers; second because even if this point could be satisfactorily settled there was still the difficulty of distinguishing the real number N from a ratio such as $N:1$. The first difficulty is not I think a difficulty of principle, though the second is. However, the most natural way of overcoming both difficulties is certainly to draw upon the quite different approach to real numbers in Chapter 3, §1, whereby real numbers are introduced precisely in terms of their position in the ordering of the rational numbers, say as least upper bounds of sets of rationals. With real numbers thus introduced in a way which gives them the appropriate relation to other numbers we can of course go on to explain the formula $a \approx_N b$ as a mere variant on $a:b = N:1$, and we can still regard this application of the real numbers as the purpose of the postulates that introduce them. That is, we could say that what makes it reasonable to adopt the theory of real numbers, and the reason why we do not think of this theory as *merely* an elegant invention of the mathematicians (like, say, hyperbolical geometries), is that this theory provides a useful way of characterizing those facts about extensive quantities which are in effect already characterizable in the theory of ratios. Since *we* apparently insist on distinguishing real numbers and ratios, it may well be argued that a direct construction of the one theory from the other will not capture our concept. But personally I am inclined to think that our insistence on distinguishing real numbers from ratios is not of such central importance that the first and simpler construction of Chapter 3 should be regarded as crucially incorrect. Rather, it suffers just from being a one-sided account of real numbers, just as a construction of natural numbers in terms, say, of ordinal adjectives alone would equally be a one-sided account of natural numbers.

List of Works Cited

BENACERRAF, P., and PUTNAM, H. (eds.) (1964). *Philosophy of mathematics: selected readings*, Prentice-Hall, Englewood Cliffs, New Jersey.

BOSTOCK, D. (1974). *Logic and arithmetic: natural numbers*, Clarendon Press, Oxford.

CAMPBELL, N. R. (1920). *Physics—the elements*, Part II. Reprinted as N. R. Campbell, *Foundations of Science*, Dover Publications, New York, 1957.

CANTOR, G. (1883). *Math. Annln.*, vol. 21, pp. 545–91.

—— (1895). *Contributions to the theory of transfinite numbers* (trans. P. E. B. Jourdain, Open Court Publishing Co., 1915). Reprinted by Dover Publications, New York, 1955; page references are to this volume.

CURRY, H. B. (1951). *Outlines of a formalist philosophy of mathematics*, Amsterdam.

DEDEKIND, R. (1872). *Stetigkeit und irrationale Zahlen* (trans. W. W. Beman, Open Court Publishing Co., 1901). Reprinted by Dover Publications, New York, 1963, in R. Dedekind, *Essays on the theory of numbers*; page references are to this volume.

—— (1888). *Was sind und was sollen die Zahlen?* Reprinted by Dover Publications, New York, 1963, in R. Dedekind *Essays on the theory of numbers*; page references are to this volume.

DUMMETT, M. (1973). *Frege, philosophy of language*, Duckworth, London.

ELLIS, B. (1966). *Basic concepts of measurement*, Cambridge University Press.

FORDER, H. G. (1927). *The foundations of Euclidean geometry*, Cambridge University Press. Reprinted by Dover Publications, New York, 1958; page references are to this volume.

FREGE, G. (1884). *Grundlagen der Arithmetik*, Breslau. Reprinted with an English translation by J. L. Austin as *Foundations of Arithmetic*, Blackwell, Oxford, 1950.

GÖDEL, K. (1944). 'Russell's mathematical logic', in Schilpp 1944. Reprinted in Benacerraf and Putnam 1964; page references are to the latter.

——(1947). 'What is Cantor's continuum hypothesis?', *Am. math. Mon.*, vol. 54, pp. 515–25. Reprinted in Benacerraf and Putnam 1964; page references are to the latter.

GOODMAN, N. (1951). *The structure of appearance*, Harvard University Press.

GOODMAN, N., and LEONARD, H. S. (1940). 'The calculus of individuals and its uses', *J. Symbol. Logic*, vol. 5, no. 2, pp. 45–55.

HEATH, T. L. (ed. and trans.) (1908). *The thirteen books of Euclid's Elements.* Reprinted by Dover Publications, New York, 1956. 3 vols.

HEATH, T. L. (ed. and trans.) (1897). *The works of Archimedes* (with supplement, 1912). Reprinted by Dover Publications, New York, n.d.

van HEIJENOORT, J. (ed.) (1967). *From Frege to Gödel, a source book in mathematical logic 1879–1931,* Harvard University Press.

JECH, T. J. (1973). *The axiom of choice,* North Holland, Amsterdam.

KLINE, M. (1972). *Mathematical thought from ancient to modern times,* New York.

NAGEL, E. (1932). 'Measurement', *Erkenntnis,* vol. 2, no. 5, pp. 313–33. Reprinted in A. Danto and S. Morganbesser (eds.), *Philosophy of science,* Meridian Books, New York, 1960.

PEANO, G. (1901). *Formulaire de mathématiques,* vol. 3. Paris.

QUINE, W. V. O. (1951). *Mathematical logic* (revised edn.), Harvard University Press.

—— (1970). *The philosophy of logic,* Prentice-Hall, Englewood Cliffs, New Jersey.

RUSSELL, B. (1903). *Principles of mathematics,* Cambridge University Press.

—— (1901), 'Mathematics and the metaphysicians', in *Mysticism and logic,* Longmans, London, 1918. Originally published as 'Recent work in the philosophy of mathematics', *International Monthly,* 1901.

—— (1919). *Introduction to mathematical philosophy,* Allen and Unwin, London.

SCHILPP, P. (ed.) (1944). *The philosophy of Bertrand Russell,* Tudor, New York.

STEVENS, S. S. (1946). 'On the theory of scales of measurement', *Science,* vol. 103, no. 2684. Reprinted in A. Danto and S. Morganbesser (eds.), *Philosophy of science,* Meridian Books, New York, 1960.

SUPPES, P. (1951). 'A set of independent axioms for extensive quantities', *Port. Math.,* vol. 10, pp. 163–72. Reprinted in P. Suppes, *Studies in the methodology and foundations of science,* Dordrecht, 1969.

TARSKI, A. (1941). *Introduction to logic* (trans. O. Helmer), Oxford University Press.

—— (1948). 'Axiomatic and algebraic aspects of two theorems on sums of cardinals', *Fundam. Math.,* vol. 35, pp. 79–104.

—— (1956). 'On the foundations of Boolean algebra', in *Logic, semantics, metamathematics* (trans. J. H. Woodger), Clarendon Press, Oxford.

WAISMANN, F. (1951). *Introduction to mathematical thinking* (trans. J. Berac), London.

WITTGENSTEIN, L. (1921). *Tractatus Logico-Philosophicus.* New translation by D. F. Pears and B. F. McGuinness, Routledge and Kegan Paul, London, 1961.

WOODGER, J. H. (1937). *The axiomatic method in biology,* Appendix E (by A. Tarski), Cambridge University Press.

Index

logic & existence? eg. 253,
261

Master argument, 202, 282

AJP rev (Dec 1975) Errol Martin, Ashgrove, Queensland (Dec 77: ANU, Canberra, RSSS)(do 1978)

Following Benacerraf, B argues prime mistake of logicism in identif'n of numbers with objects: since many different ways of doing this, none unique. Also type theory requires construction at each type-level.

B analyses numbers as quantifiers. But not single. Numbers themselves can be counted. So if quantifiers, must admit coherence of quantifying over quantifiers.

Main body of the book a 'general logic of quantifiers'.

Crit

What is presented is not what would be called a system of logic at all but a loosely-assembled mass of doctrine, explanations and appeals to intuition.

No (formal) semantics for the general logic of quantifiers is provided. So B's account of quantification remains ambiguous.

B's account of types his typed hierarchy unclear.

'logical', 'subject', 'related forms': appeals to intuition

'pure quantifiers' (may be applied to variables of any type): very ad hoc

B's theory requires '3 > 2' to be ambiguous. But is it?

Looks forward to 2nd vol.

J. Phil. rev (March 25 1976) Harold Hodes, Harvard U. (1977: Phil Dept, U of Michigan) (1978 Phil Dept, Cornell)

Legend says logicism foundered on qu 'What is logic?' and on various technical difficulties. B. :- Logic investigates relation between the structure of propositions and their truth-values. Numbers are construed as quantifiers.

Crit.
Frequent elementary confusions (of quotation with quasi-quotation, schematic constants with variables)
Obscuring informality (no interpretation of formal lang presented)
Phil. motives nebulous

Expos.
Neutral on propositions beyond minimal requisits (T or F; not both T or F; for all props there is something, viz the prop that p, that is true of and only if p) - p. 47
Neutral on whether a log. subject "ill-formed" - Hodes is the thing referred to, the expression which refers, the act of referring to it, the concept under which the thing is thought of, or what (p. 52). Subjects are characterized only by their expressions referring to (logical) objects.
A logical object iff either (i) grammatical subject-expr are in the theory taken to be logical subject-expressions. or (ii) the subject variables in the theory are taken as including that thing within their range (46)
B. : talk about components abbreviatory of talk about propositions
et "Gödel's ontological evaporation" Expands and criticizes B. on Quine

B: prop has log. subj. if exp'd with ^{taken to have}
gramm. subject-exp. which is not taken to be
eliminable

Crit. Then no subjects

" discourse apparently about numbers is really
about propositions?" "Ill-motivated"

How can we count predicates, quantifiers
and numbers, when, according to his view,
there are none?

Digression

There are at least 3 men of different height in the room

Crit

The author has not given sense to 'the same
quantifier ... applied unchanged to any
schema of the hierarchy'

Objections. "This is never explained."
"To compound the chaos, ..."
"Pure quantifiers are chimeras, born of the desire to
reap the benefits of a unified domain of discourse
without leaving a type-theory"

The sole distinction of this version of
logicism is that arithmetic comes out to be
about the traditional subject matter of logic:
propositions. The price asked of the reader, both in
patience and cash, is high.

Phil Rev rev (Jan 1976) Stanley C. Martens (Cornell)

Benac., in B's version: — "If numbers are classes, then we shd surely be able to say which classes they are, and yet this is in principle impossible" (1-2)

B. deals only with those number statements not involving plus and times.

A statement like '1 is a number' turns out to be an abbreviation for a statement about all quantifiers equivalent to "∃" ["There is at least one"]

Quantifiers are regarded not as expressions, but as components of propositions. Not objects.

(1) (x)(x loves them)   interpreted as (2) ∴
(2) Every proposition of the form "a loves them"
                                      (a)
 is true

<u>Crit</u>   B holds that he is not committed to holding that quantifiers are objects ∴ quantⁿ over quantⁿ is eliminable in favour of quantⁿ over propositions and reference to propositional forms.
B. thinks that ref. to prop. forms is eliminable

    But then quantⁿ over ordinary things is just as eliminable. So not committed to regarding ordinary things as objects.

    B's own pos. subject to Benacerraf-type argt: can define in terms of strong or weak quantifiers: no way to choose; so neither the right one.
(B. says will deal with in Vol 2)

Gödel: – no consistent formal system will have as theorems all the logical truths in Bostock's notation

Mind rev (W. Hodges)

Old claim: "we now understand that it all depends on what you count as reducing and what you count as logic ... rather boring"

On the one hand a well-known translation of Peano axioms into theorems of the simple theory of types. On the other, existential prop^ns required, not in strict sense stdents of logic.

✗ "Whether B's numbers are objects depends on what he takes his expressions to mean, and not on which of them he can periphrase away"

"Not plausible that ... could avoid treating predicates as objects just by reading '∃P' in a certain way"

"B's def of 'quantifier' is frankly metaphysical. It depends on the notion of a proposition 'containing a predicate somewhere'"...which B refuses to define
    Pure quantifiers

"The best parts of the book are also the most readable ones"